CW00970814

Artifice and Artefacts

100 Essays in Materials Science

Artifice and Artefacts

100 Essays in Materials Science

Robert W. Cahn, F.R.S.

Institute of Physics Publishing
Bristol and Philadelphia

© 1992 IOP Publishing Ltd. and Robert W. Cahn

All rights reserved. No part of this publication may be reproduced, stored in a retrieval system or transmitted in any form or by any means, electronic, mechanical, photocopying, recording or otherwise, without the prior permission of the publisher. Multiple copying is permitted in accordance with the terms of the licences issued by the Copyright Licensing Agency under the terms of its agreement with the Committee of Vice-Chancellors and Principals.

British Library Cataloguing in Publication Data

A catalogue record for this book is avaliable from the British Library

ISBN 0–7503–0152–X

Library of Congress Cataloging–in–Publication Data are available

Published by IOP Publishing Ltd., a company wholly owned by the Institute of Physics, London
IOP Publishing Ltd.,
Techno House, Redcliffe Way, Bristol BS1 6NX, UK

US Editorial Office: IOP Publishing Inc.,
The Public Ledger Building, Suite 1035, Independence Square,
Philadelphia, PA 19106

Typeset in Palatino and Photina MT by Robert W. Cahn
Printed in the UK by Cambridge University Press

Contents

Crystals, Melting and Freezing

Mechanical Properties of Materials

Order/Disorder

Glasses

Techniques of Synthesis and Investigation

Materials Aspects of Energy

Materials for Nuclear Power

Magnets

Superconductors and Superionic Conductors

Materials Conferences

Miscellaneous

Book Reviews

To Patricia

with love and gratitude

Preface

The scientific enterprise has been under repeated attack recently, and the conviction is growing among scientists that only enhanced Public Understanding of Science (the concept rates Capitals nowadays) offers any hope of stemming the hostile reaction. But it is not only the non-scientists whose understanding of science needs to be enhanced. In the face of ever expanding and diverging knowledge, we scientists are driven to retreat into our own separate domains and to peep with alarm and suspicion at what our professional neighbours are doing. Even within an established discipline such as physics or biochemistry, or indeed materials science, adherents of subspecialities are apt to keep their heads down and avoid peering perilously over the parapets. The essays in this book were written, over the years, in the hope of lowering a few of those parapets just a little. They are assembled here in the modest expectation of securing a few new readers across the scientific spectrum.

In 1968, John Maddox, then as now editor of the scientific weekly, *Nature,* invited me to become Materials Science Correspondent for the *News and Views* pages at the front of that renowned journal... anonymously at first, under my own name later on. My task was and still is 24 years later to write, at irregular intervals and in accessible prose, about diverse facets of the exceedingly broad field of materials science, which shades into physics, inorganic chemistry, crystallography, mechanics and electronics. In developing this skill I necessarily became a part-time journalist, using all available means to identify interesting subject-matter: browsing in the current literature, attending conferences and colloquia, and especially constantly conversing with fellow scientists. As the name of the column implies, not only summaries of new findings or theories but also opinion pieces have always been acceptable by *Nature.* The criterion that really matters is: Does this story interest me, and can I see a way of making it interesting to a varied readership?

Occasionally, the editorial staff would invite me to write a short piece, for the front of the journal, about a research paper that was about to appear in the back half of the journal, with a view to persuading some of the numerous readers of *News and Views* that this paper was worthy of their close attention. Some of these are also included here.

While I have always written primarily to attract the interest of materials scientists and solid–state physicists, over the years my mail has shown that other kinds of scientists have also on occasion responded to my articles. These have included chemists, various kinds of biologists, geologists, geophysicists, psychologists... to the extent that sometimes, out of sheer curiosity, I have written back to inquire about the source of the particular interest of my article for that correspondent: it generally proved to be some incidental aspect of an article, the broader interest of which had never occurred to me. It was such evidence of widespread attention, as well as interest evinced by fellow materials scientists, which motivated me to persist in the enterprise. — In selecting topics for coverage, I adopted a wide interpretation of 'materials science', ranging from oenology to the early literature of materials technology.

One test of efficacy in getting an involved 'story' across to readers is whether some of them are deceived into believing that the writer is a specialist, one who is professionally centred on that week's topic. When on occasion I was invited by editors of reputable publications to write a review paper about some topic, new to me, which I had just discussed in *Nature,* I estimated that I had 'arrived' as a science journalist!

Some articles (essays, pieces...) were also written for other journals, especially a recently launched monthly entitled *Advanced Materials,* and a cross– section of these is also included in the book, as is a selection of book reviews (I have tried to pick out those of fairly broad interest). 83 of the articles reprinted here are taken from *Nature* , 17 from other sources. Two (nos. 74 and 75) are printed here for the first time: they were written as specimen articles for an edited *Dictionary of Scientific Concepts* which was later aborted for unforeseen reasons.

The 100 articles were selected from a substantially larger total, always with the criterion of broad interest in mind. Most do require a measure of familiarity with concepts of physical science, but some (especially towards the end of the book) can be read with little or no knowledge of this sort. — I have corrected misprints, grammatical solecisms and stylistic infelicities, and a few sentences which had been edited out to save space have been restored. Some article titles have been slightly changed. Generally, I have not attempted to update the articles in the light of later research, but I have offered a few specific pieces of recent information in notes enclosed in square brackets. Apart from that, the articles are reprinted in their original form; in particular, references are cited exactly as they were at the time of first publication, following *Nature*'s changing house rules.

My particular gratitude goes to my son Andrew, who moves confidently among scientists, for eloquently urging upon me the idea of preparing this book; to the Institute of Physics for agreeing to publish it; to John Maddox for first giving me the chance to attempt scientific journalism; to hundreds of my collaborators and professional colleagues across the world for that constant stimulus which makes the profession of scientific research so irresistible to its practitioners; and above all to my wife (who is not a scientist) for her willingness to act as guineapig by reading many of these articles in order to comment on style and accessibility. Her patient support over the years, when she has detected yet another article 'coming on', has been a source of great comfort.

Robert Cahn
July 1992

Acknowledgments

I am deeply indebted to the following organizations who gave permission for articles of which they hold the copyright to be reprinted here. The exact original reference is printed under the title of each reprinted article.

Macmillan Magazines Ltd., London, in respect of articles and book reviews in *Nature* .

VCH Verlagsgesellschaft, Weinheim, Germany, in respect of articles in *Advanced Materials* and *Angewandte Chemie* .

The University of Chicago Press, Chicago, Illinois, USA, in respect of book reviews in *Technology and Culture* .

Taylor and Francis Ltd., London, in respect of a book review in *Contemporary Physics* .

Pergamon Press plc, Oxford, in respect of an article in *An Encyclopedia of Ignorance* .

The Materials Research Society, Pittsburgh, Pennsylvania, USA, in respect of an article in the *MRS Bulletin.*

Microstructure, Nanostructure and Defects

1

Limitations of Frontier Guards

Nature, **219**, 1309 (1968)

Because engineering metals are polycrystalline, their strength, ductility, formability, recrystallization, immunity from contamination — the whole gamut of their properties, in short — depend on the number and nature of the boundaries between the constituent crystals or grains. Accordingly, metallurgists have an unquenchable interest in grain boundaries, and particularly in the nature of their interaction with impurities. A brilliant paper has just been published by Ashby and Centamore[1] which is sure to stimulate much research in this field. The paper is concerned with the dragging of small oxide particles by migrating grain boundaries in copper.

The story starts with the demonstration in 1963, by Barnes and Mazey[2] at Harwell, that helium-filled bubbles in irradiated copper are able to move through the solid metal in a temperature gradient, which provides the driving force for the motion. In a sequence of electron microscope 'stills', the bubbles are seen to move just like gas bubbles rising in a glass of beer, with the essential difference that the copper is and remains solid. Shewmon[3] concluded that metal diffuses along the bubble wall in counter-flow to the direction in which the bubble is migrating.

Ashby, working at Harvard, has for some time been interested in the possibility that solid inclusions in metals, which in a sense are merely stuffed bubbles, may also be capable of migrating through the solid metallic matrix. This would be of distinct practical significance, because dispersed particles in metals are of ever-growing importance in conferring high strength, especially at high temperatures. A part of the importance of particles in strengthening metals comes from their ability to inhibit grain boundary migration. Thus extruded 'thoria-disperse nickel' owes its strength largely to a high concentration of dislocations locked in position by the thoria particles, and if new grains could grow into this deformed structure, the dislocations would disappear and the metal would be seriously weakened. Hence the practical interest in the possibility that a moving boundary may be able to drag along solid particles.

Ashby and Centamore have carried out a series of beautifully clear and conclusive experiments on highly alloyed copper samples which, by internal oxidation, have been provided with dispersions of B_2O_3, GeO_2, SiO_2 or Al_2O_3. Grain boundaries were caused to migrate in these samples at about 1 000°C under a driving force just insufficient to allow them to break free from the dispersions. Thus the boundaries could move only by dragging the particles along, and if they did this, a band of copper swept clear of particles was left behind. The boundaries were found to be carrying all the particles which had been swept up; there was no coalescence. The width of the band was measured for different temperatures, particle sizes and alloy systems. The ease of dragging diminished in the four systems in the order quoted above — B_2O_3 particles moved readily while Al_2O_3 particles could not move at all.

Near 1 000°C, B_2O_3, GeO_2 and SiO_2 are all viscous fluids, while Al_2O_3 is not fluid. The authors conclude that viscous fluid inclusions migrate like bubbles, and comparison with theory confirms that at higher temperatures the diffusion of copper along the matrix/inclusion interface is the controlling mechanism. At lower temperatures, the rate of dragging declines steeply, and the authors postulate that diffusion of copper through the particles plays an essential part. This, of course, explains why Al_2O_3 particles are not mobile at all — diffusion through them is impossible.

In passing, the authors conclude that a coherent inclusion, which has a lattice parallel to and continuous with that of the matrix, should be immobile because of the difficulty of putting vacancies into the interface. This prediction is likely to be correct but will be difficult to test because it is known that coherent particles can be dissolved in the matrix at an advancing grain boundary and then reprecipitate in coarser form once the boundary has passed. This has been demonstrated in a nickel alloy containing coherent γ' (Ni_3Al) inclusions[4] and in copper containing coherent cobalt inclusions[5].

Another recent paper has emphasized a further way in which alloy constituents may affect the behaviour of grain boundaries. Dorward and Kirkaldy[6] have demonstrated that the tendency of solute to concentrate at grain boundaries causes the small solubility of copper in single crystal silicon to increase by a large factor when grain boundaries are present; for example, at 650°C a 350-fold increase was found. The existence of this potentially serious source of error was first established 20 years ago by Voce and Hallowes[7], who showed that the (small) solubility of bismuth in copper in fact depends on the grain size, but this work seems to have been forgotten. Solute segregation at grain boundaries is also of practical importance, for dissolved impurities exert a drag on a moving grain boundary just as discrete particles do, and in recent years a body of experiment has been accumulated to describe the role of this drag in determining recrystallization kinetics and also the nature of the preferred orientation created during recrystallization. These studies have thrown up some theoretical puzzles.

1. M.F. Ashby and R.M.A. Centamore, *Acta metall.* **16**, 1081 (1968).
2. R.S. Barnes and D.J. Mazey, *Proc. R. Soc. London, Ser. A* **377**, 47 (1963).
3. P.G. Shewmon, *Trans. Metall. Soc. AIME* **230**, 1134 (1964).
4. F. Haessner, E. Hornbogen and M. Mukherjee, *Z. Metallkde.* **57**, 270 (1966).
5. L.E. Tanner and I.S. Servi, *Mater. Sci. Eng.* **1**, 153 (1966).
6. R.C. Dorward and J.S. Kirkaldy, *J. Mater. Sci.* **3**, 502 (1968).
7. E. Voce and A.P.C. Hallowes, *J. Inst. Metals* **73**, 323 (1946-1947).

2

Passage of Ions through Crystals

Nature, **220** , 960 (1968)

Electron microscopy of thin crystal films has become one of the materials scientist's indispensable tools. The technique owes much of its usefulness to the fact that electrons moving through a crystal in particular directions are anomalously absorbed and thus contrast patterns are observed which can be related to the local orientation of a crystal or to its contained imperfections.

Analogously, the passage of heavy positive ions through a crystal can also give rise to anomalous absorption, and this has in the past few years been the subject of a rapidly burgeoning research effort. One end product of this work, a useful "proton microscope" for studying the symmetry of the transmission pattern of protons passing through a crystal, may soon be commercially available[1]. Quite apart from this, however, the potentialities of this field of research deserve to be widely known, and a recent comprehensive review article[2] is therefore timely. The principal related features which have been studied include channelling and focused collision sequences, and these in turn have been closely connected with studies of two phenomena of technological importance, sputtering and ion implantation.

An ion moving through a crystal is said to be channelled when it moves accurately parallel to a low-index crystal direction or crystal plane (in the latter instance, its azimuth is immaterial). Professor Thompson's review is primarily concerned with this phenomenon. Channelled ions are much less rapidly absorbed than ions moving in arbitrary directions, because all their collisions with lattice ions are glancing ones. Much of what is known about the process stems from measurements of penetration of monoenergetic ions in different directions and of the energy distribution of the various transmitted ions. Most of these studies have been concerned with heavy ions — usually noble gas ions — but last month a very detailed study appeared on the channelling of protons through various kinds of simple crystal structures[3]. The anisotropy of transmission was studied photographically and the energy spectra by means of a semiconductor detector, but the two types of information were also to some degree combined by a remarkably elegant technique due to the four first named authors[4], involving the use of multi-layered colour film into which ions of different energies penetrate to different distances, so that the colour of the developed image is an indication of ion energy. (This technique was originally developed to help in studying the injection of dopant ions into semiconductor crystals.)

The new study has shown particularly clearly that channelling is always accompanied by its converse, blockage. A proportion of incident protons, instead of being channelled, are scattered through large angles, giving a diffuse transmission background. This background pattern is marked by a series of absorption lines, rather like Kikuchi lines in electron diffraction, which represent directions in which the once-scattered ions might have been expected to be channelled. These paradoxical protons are not channelled because they were all scattered by lattice ions, and so, while they were moving in the right directions, their motion started in a lattice plane and not between a pair of adjacent lattice planes, as is needed for proper channelling. A proton moving in a lattice plane is especially likely to be scattered a second time, and so the absorption patterns result. Dearnaley and his colleagues have studied the intriguing transitions from anomalous transmission to anomalous absorption as the direction of incidence of the protons is gradually altered from an exact crystallographic vector, and from this have deduced information about the existence of a potential barrier between axial and planar channelling. The phenomena have all been interpreted in terms of a simple physical model for proton-ion collisions.

1. R.G. Livesey and G. Butcher, paper presented to the fourth International Vacuum Congress, Manchester, 1968.
2. M.W. Thompson, *Contemporary Physics* **9**, 375 (1968).
3. G. Dearnaley, I. V. Mitchell, R. S. Nelson, B. W. Farmery and M. W. Thompson, *Phil. Mag.* **18**, 985 (1968).
4. G. Dearnaley, I.V. Mitchell, R.S. Nelson and B.W. Farmery, *J. Mater. Sci.* **2**, 1971 (1967).

3

Microstructure Analysed

Nature, **224**, 535 (1969)

Measuring the grain size of a piece of metal may not be as straightforward as it seems. The routine method for determining the grain size of a metal, alloy or ceramic is to superimpose a series of regular honeycomb grids of different coarseness on a magnified image of a polished and etched section until a match is obtained. But the real grain structure, unlike the grids, is subject to many kinds of statistical scatter. The maximum diameter, ratio of maximum to minimum diameters, number of sides, mean diameter, mean length of sides and the form of distribution of diameters can all be expected to vary from one sample to the next; yet some of these variables, such as the form of distribution of diameters, may behave more regularly than others. The experimental methodology and mathematical superstructure involved in investigating such variables constitute the little known science of stereology. Metallurgists often call it quantitative metallography, but the more general name is preferred by biologists concerned with the quantitative geometrical study of cells and their microconstituents. An intriguing set of stereological regularities has been published recently by Aboav and Langdon[1]. They examined the grain structure seen on sections of a batch of transparent (i.e., pore-free) magnesia, of average isotropic grain size ≈10 mm. Their conclusions, based on the statistical analysis of some 10 000 grains, can be summarized as follows. In a plane section of a polycrystal: (a) the average diameter of grains with n sides is proportional to $(n-2)$; (b) the distribution of the lengths of grain sides is of the same form as that of the grain diameters, but has a slightly larger spread; (c) the form of the distribution of diameters of grains with n sides is independent of n; and (d) the proportion of grains with n sides (n varies from 3 to 12) can be expressed as a Gaussian function of $\log n$, or of $(n-2)^{1/2}$.

Aboav and Langdon did not attempt to analyse these findings in topological terms; their paper is purely empirical. Their last conclusion, combined with an earlier finding, by P.A. Beck, that the proportion of grains (in aluminium) with *n* sides does not vary during grain growth, raises the interesting question of whether the invariance of the grain morphology as expressed by the fourth conclusion can be interpreted in terms of the unit processes that govern grain growth; these processes, at the lines and points of intersection of trios and quartets of grains respectively, are well understood since M. Hillert's analysis of grain growth. Such an interpretation has yet to be attempted.

Aboav and Langdon's analysis is comparatively straightforward. Many stereological analyses, however, are more complex; even though the actual measurement may be simple (such as, for example, a count of the density of intercepts by phase boundaries along a randomly placed line), the underlying theory is elaborate. Moreover, the problems become intrinsically more complicated when more than one phase is present in a microstructure: one may then need to know, for example, the volume fraction of the second phase, the total area of interface between the phases per unit volume, or the degree of anisotropy of the disperse phase. Again, for the study of certain forms of precipitate coarsening in alloy systems, it is necessary to know the statistical distribution of true diameters of a population of spherical precipitates of various sizes, to be deduced from measurements made on cross-sections. Here one must allow for the fact that most sections of spheres are non-central and therefore appear smaller than the sphere itself. A subtle problem recently settled with the help of a computer was the question of whether a large array of irradiation damage spots in a foil of irradiated copper was distributed at random or preferentially in linear arrays[2].

1. D. Aboav and T.G. Langdon, *Metallography* **2**, 171 (1969).
2. F.J. Minter and A.J.E. Foreman, *J. Mater. Sci.* **4**, 218 (1969).

4

Anisotropy of Surface Tension

Nature, **224**, 948 (1969)

It is well established that the surface energy of a monocrystalline solid is a function of the orientation of the plane on which the measurement is made: indirectly, this anisotropy is the origin of the faceting of crystals and indeed of the fact that some crystals cleave. What is much less familiar is the notion that surface tension (which is intimately related to surface energy) can be anisotropic with respect to direction in a crystalline plane. Unambiguous evidence of this form of anisotropy has been published recently by Asselmeyer and Mecke[1]. They prepared cleavage surfaces of artificially grown sodium chloride crystals and measured surface tension by means of the classic sessile drop technique, using mercury as the reference liquid. The contact angle was measured as a function of azimuth in the cleavage plane. Exceptional care was taken to test the technique and identify sources of error, and there can be no doubt that the results are significant.

It turned out that in certain circumstances the plot of contact angle against azimuth over the 45° range between [100] and [110] showed no fewer than six precisely reproducible maxima. The contact angles were very sensitive to humidity and reproducible anisotropy was found only for low humidities. Annealing of the cleaved samples destroyed all trace of anisotropy. The anisotropy is thus not an intrinsic property of a smooth crystallographic surface: on the contrary, Asselmeyer and Mecke attribute it to a postulated roughness of the cleavage "plane", which they take to be largely made up of microfacets of rational planes other than (100). The anisotropy of surface tension can then (following a calculation due to Kuznetsov) be related to the specific energy and orientation of these facets. Asselmeyer and Mecke adduce several independent lines of evidence for the existence of the microfacets, including their own studies of the adsorption isotherms of water on cleaved sodium chloride. They might have added the extensive body of experimental results on the decoration of cleavage planes of sodium chloride by means of evaporated gold; these studies (at CSIRO in Australia) have shown the vital importance of the microfacet structure for the occurrence of epitaxial deposition of metals on salt substrates. The loss of anisotropy following annealing or in the presence of excessive humidity is attributable to the smoothing out of the microfacets. It would be interesting to establish a direct correlation between epitaxial efficiency and anisotropy of cleavage surfaces treated in various ways.

Asselmeyer and Mecke's study was prompted by their own recent work[2] on the velocity of propagation of cleavage cracks in sodium chloride, in which they found this velocity to be markedly anisotropic. Because the spread of cleavage cracks is dominated by the surface energy of the newly created crack surface, their recent findings serve to interpret these earlier results, which should indeed be of wider interest to those interested in fracture mechanics. Their results raise a rather intriguing problem: Griffith's equation for the stability of a cleavage crack includes the surface energy, a quantity which cannot be anisotropic within a plane. The surface tension, which can be and is anisotropic within a plane, has the wrong dimensions for Griffith's equation, yet plainly it is the right quantity to use if the anisotropy of cleavage velocities is to be explained.

1. Asselmeyer and Mecke, Z. *Angew. Phys.* **28**, 53 (1969).
2. Asselmeyer and Mecke, *Naturwiss.* **55**, 129 (1968).

5

Topology of Crystal Grains

Nature, **250**, 702 (1974)

It is one of the paradoxes of the history of science that the rigorous quantitative treatment of the behaviour of large populations of molecules preceded by many years the similarly rigorous description of the individual molecules themselves; yet the kinetic theory of gases implies the extraction of orderly and predictable behaviour from myriads of random motions and collision, whereas molecules are all identical. The resolution of the paradox lies in the fact that in the kinetic theory, molecules are treated virtually as independent, featureless particles, and therein lies the tractability of the whole approach. A million particles spell simplicity; one molecule spells complexity.

With crystals, the sequence is reversed. A large number of crystal structures had been accurately determined before the collective behaviour of populations of crystals began to be understood. The metallurgist recognises two distinct forms of such collective behaviour: first, there is 'Ostwald ripening', a poetically metaphorical term for the Matthew principle applied to a population of spherical crystallites of varying radii dispersed in a matrix; the larger ones grow at the expense of the smaller, essentially because of a radius dependence of the solubility of the particles in the matrix. (Water droplets in a cloud behave in an essentially similar way.) The rigorous treatment of this very important metallurgical process is now well understood.

The second form of collective behaviour is grain growth. A piece of a pure metal consists of a population of crystal grains all sharing the same crystal structure but of different sizes, shapes and orientations. The grains grow from independent nuclei and as they impinge, grain boundaries are formed. The interfaces may be plane or curved, and bear no relation to the regular lattice arrangement of atoms: "There's no art to read the grain's construction in the face". Here, also, the Matthew principle operates: if a piece of metal is heated, the larger grains grow at the expense of the smaller through the migration of grain boundaries, and the average grain size progressively increases.

Grain growth has been studied micrographically for many years, both for its intrinsic interest and because of its practical influence on metal working, on the behaviour of nuclear fuel elements, in powder metallurgy and elsewhere. It has long been recognised that the mean grain diameter varies withtime according to $4D = kt^{n}$, where n falls in the range 1/3–1/2. The problem is to understand why. It is a conceptually more difficult problem than the kinetic theory of gases, because crystal grains are not independent and separate particles but connected polyhedra. When one grain changes shape and size, the neighbours of necessity change too. An assembly of grains is like a nuclear reactor: a change in neutron absorption in one small corner soon leads to a change in neutron flux throughout the reactor. Similarly, an instability at one corner of one grain quickly spreads through the population.

The key to an understanding of the problem is its topology. A population of polyhedral grains can be characterised by the number of grains, faces and vertices, and by the relations between these quantities. In particular, interest attaches to the mean number of faces per grain and the distribution of this number among the grain population. Topological principles were first applied, to assess the stability of a grain population, by C. S. Smith in 1952. He recognised that grain growth stems from disequilibrium at edges where three grains meet and at four-grain corners. Unless three boundaries meeting at an edge are mutually inclined at 120°, the edge must move and the boundaries become curved. The curvature itself introduces instability; indeed, curvature and non-equilibrium edge and corner configurations are the fuels of grain growth. The nearest approach to equilibrium obtains in a population of fourteen-sided polyhedra of a particular shape.

In a brilliant paper, Rhines, Craig and Dehoff[1] apply topological analysis to grain growth in a highly illuminating way. Their article is based upon an experimental study of successive stages of grain growth in aluminium, using the technique of serial sectioning; the technique depends on a thorough mastery of the theory of quantitative metallography which allows two-dimensional measurements in the microscope to be converted into reliable three-dimensional geometrical statistics. In their laboratory at the University of Florida, the authors have honed this technique to a fine cutting edge. They are able to determine the total grain boundary area, mean boundary curvature, distribution of the number of faces per grain, as well as the true mean volume per grain, as a function of time.

By combining experiment with topological analysis, Rhines and his coworkers[1] were able to introduce the concept of a "structural gradient", which is essentially a measure of the mean deviation from equilibrium of the entire grain structure. It is equal to the product of the total mean curvature and the surface area per grain. This structural gradient determines the rate of steady-state grain growth, and remains unchanged because the form of the distribution of different grain shapes remains unaltered as grain growth proceeds. All this is made topologically clear by the authors. They also show in a simple and elegant fashion that the number of grains eliminated when a grain boundary sweeps through unit volume of material is a constant, independent of the mean grain size. This very important theoretical result will have a number of important applications in materials science (for instance, in connection with sintering of powders, in which grain growth plays a crucial part).

By putting the various parts of the analysis together, the authors find both theoretically and experimentally that grain growth follows the law $\Delta D = kt^{1/3}$, when the 'diameter' D is expressed as the cube root of the mean true grain volume. The apparent grain diameter derived from intercept analysis on a single two-dimensional section gives an index rather higher than 1/3, which indicates that good quantitative metallography requires more expertise than most investigators possess. The authors point out that "it has not been required, in developing the foregoing rate law, to introduce any arbitrarily adjustable parameter, as has been done in the usual expression of grain growth kinetics". Theirs is an impressive achievement.

A recent attempt by Louat[2] to apply purely statistical principles to grain growth is an instance of the type of analysis at which the Florida authors, by implication, cock a sceptical eyebrow. This is not to say that the statistical approach has no value: for instance, Weaire[3] shows how statistical arguments can be combined with topological ones to assess whether a distribution of grain shapes deviates locally from randomness.

Another recent article of striking originality on a related theme is by Morrall and Ashby[4] who consider an assembly of fourteen-sided polyhedra in near equilibrium, as per Smith's specification, and then introduce a number of thirteen or fifteen-sided grains, or other more serious 'grain defects'. The grain structure is then denoted by joining the centres of all neighbouring grains, through their common boundaries, by a network of lines. These lines form a lattice ('lattice graph') with dislocations wherever there is a grain defect. These dislocations can move conservatively (using the term in the special sense of dislocation theory); this corresponds to grain displacement of the type found in superplasticity, without change in the number of grains. Dislocations can also move by climb, which implies the disappearance of some grains in 'real space' and therefore corresponds to grain growth. The authors build on an earlier analysis by Hillert of grain growth in two dimensions, where a set of perfect hexagons is disturbed by some rogue pentagons, and the like. For three dimensions, they relate, as did Hillert for two dimensions, the rate of grain growth to defect density, that is, to the density of dislocations in the lattice graph. They predict that if the defect density is constant, $\Delta D = kt^{1/2}$, while the index <1/2 if the defect density decreases during grain growth.

Morrall and Ashby's analysis represents an interesting link between the purely topological and purely statistical approaches. Since the Florida authors showed experimentally that the grain shape distribution is invariant with time (that is, the grain defect population is invariant too) Morrall and Ashby's analysis predicts $\Delta D = kt^{1/2}$, which is inconsistent with the experimental findings of the Florida group. The problem of grain growth has plenty of mileage left for those investigators who can master the requisite degree of subtlety.

1. Rhines, Craig and Dehoff, *Metall. Trans.* **5**, 413 (1974).
2. Louat, *Acta metall.* **22**, 721 (1974).
3. Weaire, *Metallography* **7**, 157 (1974).
4. Morrall and Ashby, *Acta metall.* **22**, 567 (1974).

6

Harmony of the Spheres

Nature, **251**, 103 (1974)

Very gingerly, the proponents of fibre reinforcement are beginning to approach the serious possibility of using fibre-reinforced materials for service at high temperatures – for instance, in gas turbine components. One central difficulty about this is that the metallic matrix may chemically attack the fibres; for example, tungsten fibres interact chemically in a complicated way with a nickel matrix. The other tricky problem is to keep the shapes of the fibres inviolate. It is not too difficult to generate a regularly spaced array of long, thin fibres in a suitable matrix; much research has been done on the directional growth of monocrystals of appropriate binary and ternary alloys, consisting of aligned fibres (or plates) in a matrix. The trouble is that a long thin cylindrical fibre is not in geometrical equilibrum with its matrix.

In a slowly grown eutectic, the total quantity of fibrous phase will be in equilibrium with the total quantity of matrix: the disequilibrium is purely a matter of geometry. This has long been known to metallurgists familiar with ordinary mild steel. The pearlite ($Fe+Fe_3C$) eutectoid lamellae, if sufficiently heated, turn into arrays of fine Fe_3C spheres. The process, spheroidisation, is driven purely by the tendency to minimise the total area of Fe/Fe_3C interface. The same can happen to a long fibre in a matrix.

As so often in physical science, one of the key ideas comes from Lord Rayleigh. In 1878 he showed that a cylindrical jet of water tends to instability: it is apt to pinch off at intervals of about eight diameters. Recently McLean, of the National Physical Laboratory showed[1] that, correspondingly, cylindrical inclusions of lead in aluminium split into separate short cylinders if the aspect ratio (length/diameter) exceeds 8. Each cylinder then gathers up into a sphere. The relevance to the stability of a population of fibres is manifest. Although aluminium-lead is no candidate for high temperature use, it is a useful model system.

With the collaboration of Loveday, McLean[2] has now taken his investigation of Al/Pb alloys a long step forward. The main improvement is one of technique: instead of using X-ray microradiography between anneals, the investigators examined their foils in a 1 MeV electron microscope fitted with a heating stage. With the aid of cinematography, and exploiting the high penetrating power of the energetic electron beam, the investigators were able to trace in detail the stages of spheroidisation and make accurate kinetic measurements. In addition to the advantages of *in situ* micrography, they also achieved a much higher resolution by this technique.

The lead inclusions in the Al + 5 wt% Pb alloy were turned into cylinders by mechanical swaging (there is no suitable eutectic) and foils cut from this material were heated in the microscope to above the melting temperature of the lead. The spheroidisation process is governed by diffusion of matrix atoms through the cylinders; since the lead was molten, diffusion was rapid and so was the spheroidisation. The cylinders either broke up into shorter ones which then turned into spheres, or short cylinders spheroidised directly; Rayleigh's criterion was approximately obeyed. The detailed observations allowed the diffusivity of aluminium in molten lead to be determined for a range of temperatures.

McLean and Loveday were also able to examine the merging of individual spheres once they had formed. The lead spheres behaved just like the bubbles of helium in nuclear fuels which have been extensively investigated in recent years, migrating under the driving force merely of a temperature gradient. The kinetics of merging of large spheres (diameter $\approx$1 μm) was studied and found, again, to be controlled by volume diffusion of aluminium in liquid lead. These large spheres moved at a speed, independent of their size, which was consistent with the estimated temperature gradient in the microscope hot stage. Spheres with diameter less than 0.8 μm moved more slowly, the smaller the sphere; this is proof positive that the motion of these small spheres was controlled by interface energy and not by volume diffusion. Very small spheres, less than 0.12 μm in diameter, did not move at all. This immobility is a new phenomenon and its origin is not clear. The observations showed the merging of pairs of large spheres when one catches up another, and the kinetics of this process was consistent with the theory developed by earlier investigators of bubbles in nuclear fuels. It seems that bubbles/spheres behave in the same way, irrespective of whether they are filled with gas or with molten metal. McLean and Loveday point out that their observations have practical implications (quite apart from their relevance to the stability of composites). Free-machining alloys such as leaded brass contain small particles of soft metal in a hard alloy matrix, and it should now be possible to develop a prescription for a combined mechanical/thermal treatment to produce the form of spherical dispersion with the best machining characteristics.

1. McLean, *Phil. Mag.* **27**, 1253 (1973).
2. McLean and Loveday, *J. Mater. Sci.* **9**, 1104 (1974).

7

Vacancies in Nickel-Aluminium and Other Alloys

Nature, **279**, 579 (1979)

In 1937 the metallurgist Abraham Taylor made a celebrated study of the structure of nickel--aluminium alloys; this called for great experimental precision, and in 1972 he returned to the problem and examined it with even greater scruple (Taylor and Doyle[1]). This second study may be said to have launched a new wave of research on the Ni--Al system and its analogues, even though few of the most recent protagonists seem to be aware of Taylor's second study.

Taylor's experiments centred on the alloy β-NiAl, which adopts the very simple B2 (CsCl-type) structure: nickel atoms at cube corners, aluminium at cube centres. Taylor measured macroscopic densities and lattice parameters as a function of composition, either side of the 50/50 composition: by putting together the two sets of measurements, he was able to deduce the destinations of excess nickel or aluminium atoms. On the nickel-rich side, the extra nickel atoms substitute in the usual way for aluminium atoms on the aluminium sublattice. The aluminium-rich side, however, is quite different: no aluminium substitutes on the nickel sub-lattice; instead, nickel atoms disappear from the nickel sublattice, leaving nickel vacancies. For instance, at 45at. % Ni, 18 % of the nickel sites are vacant, all of the aluminium sites are filled. Such vacancy populations, determined by composition and not temperature, are distinguished as constitutional vacancies.

Several other studies showed that stoichiometric β-NiAl quenched from a high temperature (as opposed to that slowly cooled) contained a high concentration of thermal vacancies; the most recently cited figure is 1.08% of vacancies at 1 600°C. This is a very much larger thermal vacancy concentration than is found in other metals or alloys, even just below the melting temperature; so large that, on cooling, the vacancies will separate out into a population of voids visible in the electron microscope (Epperson[2]). 50/50 NiAl which contains such vacancies, all on the nickel sublattice, must also contain substitutional defects – that is, some nickel atoms in the aluminium sublattice, also called nickel antistructure atoms – to preserve the overall chemical composition: specifically, two vacancies must be accompanied by one substitutional defect. Such a trio of linked defects is now termed a triple defect and much recent research has centred on this entity.

In recent years, investigators in the US, England andd Germany began to look for parallels to this curious behaviour in alloy systems isomorphous with β--NiAl, and attention soon centred on NiGa and CoGa. It became clear that these two phases always contain triple defects, even when they have been cooled slowly to eliminate thermal vacancies as far as possible. Though there has been some confusion in the literature on this point, it is now adequately clear that the triple defects in these alloys at 50/50 composition are indeed wholly of thermal, not constitutional origin; it is impossible to eliminate all thermal defects, even by very slow cooling.

A study of NiGa by Donaldson and Rawlings[3] again made use of measured densities and lattice parameters to assess just how the concentration of nickel vacancies and gallium antistructure atoms varies with composition. To obtain these values, they developed an iterative method of calculation, starting out with an arbitrary assumption about the number of triple defects at stoichiometry, and refining this value in cycles until experimental densities and lattice parameters at a range of compositions could be matched. (To do this, it is necessary to assume a value for the relaxed volume of a nickel vacancy, which has to be based on theoretical estimates, and comparison of some recent papers shows that this is the weak link in the chain of theoretical deduction.) Donaldson and Rawlings found just over 3 % of vacancies and half that number of antistructure atoms in 50/50 NiGa. At 47 % Ni, there are 7% vacancies and 0.2 % antistructure atoms, while at 54 % Ni, the defect numbers are 1% and 4.5% respectively. The triple defect itself (a 2/1 ratio of the two kinds of defect) only exists at stoichiometry. Ho *et al.*[4] obtained rather similar values, though they do not match Donaldson's and Rawlings' figures in every particular.

Berner[5,6] has found closely similar values for defect densities in CoGa.

The computational problem of determining defect concentrations from experimental data when both defect types are present has now been eased by an ingenious application of statistical thermodynamics to the problem. Edelin[7] calculated the relations between composition, defect concentrations and defect formation energies for NiGa, using merely the hypothesis that the formation energies for Ga vacancies and antistructure atoms on the Ni (or Co) sublattice are so much higher than the energies for Ni (or Co) vacancies and antistructure atoms on the Ga sublattice that the former two defects never appear in detectable numbers. Edelin showed that the variation of concentration of the two defect types, both as a function of temperature and as a function of composition, was determined by a single composite activation energy, Q , and Q can be determined directly by fitting, for a range of compositions, experimental values of density and lattice parameter to his theoretical equations. Further, it appears that when fitting is done in this way, the relaxed volumes of vacancies and of antistructure atoms can be deduced from the measurements and do not need to be used as assumed input values. (This claim seems so remarkable that it deserves critical appraisal by a suitably qualified theoretician: it would seem to be the first claimed method for deductively obtaining relaxed defect volumes from experimental data.)

The claim by Edelin to be able to measure the relaxed volumes of defect sites is of particular interest because it has been established, from diffuse X-ray scattering measurements made at Stuttgart, that the lattice near a vacancy in NiAl or CoGa is severely distorted in a [111] direction. This distortion is of the same type as is found in a $\beta \rightarrow \omega$ phase transformation. Such distortions must surely affect the relaxed vacancy volume (Ortiz and Epperson[8], for NiAl; Kirchgartner and Gerold[9], for CoGa).

The composite activation energy, Q, found by Edelin for NiGa is 0.29-0.32eV per atom and for CoGa, 0.265 eVper atom. These very low values account for the high thermal defect concentrations found even at stoichiometry in these two alloys. Edelin's method has not been used for NiAl (no accurate measurements of thermal defect concentrations at stoichiometry being available), but Q can be estimated from measurements of nickel diffusionin NiAl at a range of compositions (Hancock and McDonnell [10]). On the low-nickel side, it is assumed that the measured activation energy for diffusion is purely that for vacancy migration (because there is an abundance of constitutional vacancies available); then it is possible to compute the formation energy for thermal vacancies by subtracting the migration energy from the measured activation energy for diffusion a tstoichiometry (where no constitutional vacancies exist). This approach gives 0.70 eV per atom, corresponding to themuch smaller thermal vacancy concentration in this alloy. This much higher Q value also accounts, by way of Edelin's theory, for the virtual absence of aluminium antistructure atoms on the nickel-poor side in this alloy series, as contrasted with NiGa.

There have been several attempts to bypass the density/lattice parameter approach, which makes severe demands on experimental accuracy. (One difficult problem is to get precise densities in view of the presence of casting porosity in ingots; Taylor andDoyle overcame this by high-pressure annealing of ingots, and other investigators have sought to bypass it by using dilatometry instead of Archimedean weighing.) An alternative approach is to measure thermodynamic activities at high temperature and to relate such activities to defect concentrations. Atheory to make this possible has beenconstructed by Neumann, Chang andLee[11] and has been applied to NiGa[12]. The vacancy concentrations in NiGa alloys deduced from activity values are considerably lower than those obtained by the conventional method with quenched samples, but a very recent paper (Neumann and Chang[13]) shows that if allowance is made for loss of vacancies during quenching, then agreement is better. Nevertheless, it does seem that there must be some residual error in Neumann's statistical thermodynamics and one possible source of such error is the tacit assumption that vacancies and antistructure atoms are randomly disposed. Delavignette, Richel and Amelinckx[14] have adduced electron-microscopical evidence pointing to such order among the vacancies in hypostoichiometric NiAl, and Reynaud[15] has evidence suggesting ordered antistructure atoms in hyperstoichiometric NiAl. Any short-range order among defects would have a considerable effect on thermodynamic activities. The recently observed anisotropic distortion around vacancies, already cited, with its associated stress field, could also affect activities. The activity method is therefore suspect, but it must be admitted that substantial order or lattice distortions could introduce errors into the density/lattice parameter method as well.

Surprisingly little attention has been paid to the question why β–NiAl, NiGa and CoGa form defect structures at all. For instance, Massalski and Mizutani, in their survey of electronic structures of Hume-Rothery phases[16], make no reference to these phases. Taylor and Doyle[1] point out that the electron/atom ratio remains strictly constant at 3/2 from 50/50 NiAl to the phase boundary on the aluminium-rich side (where there are only 1.75 atomsper cell), and they assume this feature to govern the introduction of vacancies.The question then arises why other isomorphous alloy series such as Ni-Zn or Co-Be do not behave in this way. Machlin[17] has made an ingenious approach to this question by suggesting that the self-energy of the B (non-transition) component may be lowered by a reduction in the number of bonds between B and transition metal atoms, resulting from the presence of transition metal vacancies adjacent to some B atoms. He is able to show that on this hypothesis, constitutional vacancies should be stabilised only for trivalent B metals, such as Al or Ga, and this is in accord with observation. There is scope for further theoretical treatment of the reasons for the formation of constitutional vacancies.

Clearly, NiGa and CoGa in particular are ideally suited as test materials in which the predictions of precise statistical thermodynamics applied to defect populations can be compared with experimental observations. They are probably the best metallic materials for such comparisons, free of the complications arising from the need for charge neutrality which are unavoidable with ionic crystals.

1. Taylor and Doyle, *J. appl. Cryst.* **5**, 201 (1972).
2. Epperson *et al.*, Phil. Mag. A **38**, 529 (1978).
3. Donaldson and Rawlings, *Acta metall.* **24**, 811 (1976).
4. Ho *et al.*, *Scripta metall.* **11**, 1159 (1977).
5. Berner, Dissertation, Stuttgart, 1976.
6. Berner *et al.*, *J. Phys. Chem. Solids* **36**, 221 (1977).
7. Edelin, *Acta metall.* **27**, 1455 (1979).
8. Ortiz and Epperson, *Scripta metall.* **13**, 237 (1979).
9. Kirchgartner and Gerold, *J. appl. Cryst.* **11**, 153 (1978).
10. Hancock and McDonnell, *Phys. stat. sol.* **A4**, 143 (1971).
11. Neumann, Chang and Lee, *Acta metall.* 24, 593 (1976).
12. Neumann, *Scripta metall.* 11, 969 (1977).
13. Neumann and Chang, Z. *Metallkde.* **70**, 118 (1979).
14. Delavignette, Richel and Amelinckx, *Phys. stat. sol. (a)* **13**, 545 (1972).
15. Reynaud, *J. Appl. Cryst.* **9**, 263 (1976).
16. Massalski and Mizutani, *Prog. Mater. Sci.* **22**, parts 3/4 (1978).
17. Machlin, *Scripta metall.* **13**, 123 (1979).

8

Penny Plain, Tuppence Coloured

Nature , **320**, 304 (1986)

A flourishing community of theoreticians seeks to clarify processes in crystals — phase transformations, critical phenomena, radiation damage, melting and vitrification— by simulating them in powerful computers. The latest attempts to model crystals realistically are reported in two papers elsewhere in this issue[1,2].

Physical modelling of crystals began in the late 1940s with the two-dimensional soap-bubble raft of Bragg, Nye and Lomer. Given the right bubble radius, such rafts reproduced well the interatomic force law typical for metals, and were used to simulate the motion of dislocations under shear stress. They were also used to model metallic glasses, but for this purpose, two populations of bubbles of different radii were essential to inhibit 'crystallization' [3]. Such rafts have been well used, notably by Argon[4], to interpret mechanical properties of metallic glasses.

In the 1960s, Bernal produced his influential liquid model with monosized steel balls kneaded in a football bladder which was later extended to imitate metallic glass structures. In contrast to the bubble raft, Bernal's model suffered from the fact that the balls were hard and could not imitate the 'soft' interatomic force laws appropriate for alloys: this inadequacy was overcome, both literally and metaphorically, by the introduction of software to allow relaxation of a Bernal model via computer simulation.

More recently, a new family of 'macroparticulate' crystals has appeared in the form of colloidal crystals[5], regular assemblies of equal submicron-sized spheres, typically of a polymer or of silica. These crystals were not intended as models: their study grew from observations of natural macroparticulate crystals — the crystalline viruses and opals[6] (which last consists of regular arrays of monosized amorphous silica spheres). The first artificial macroparticulate crystals to be made, gem-quality opals, were made by gravity-induced sedimentation of colloidal silica suspensions, by Pierre Gilson at his factory in northern France[7] and have been in commercial production since 1974. How these opals are stabilized after sedimentation (sintering or low-temperature adhesion) is a commercial secret.

The study of silica and polymeric macroparticulate crystals has accelerated rapidly since then (although it is doubtful whether many of the physicists concerned know of Gilson's early triumph). The use of macroparticulate crystal assemblies of ceramic microspheres as 'green' precursors for sintering has now been proposed and a promising start has been made with silica[8] (as discussed in these columns[9]).

One of the papers in this issue[1] follows the colloidal approach to crystal simulation: Pusey and van Megen used mono-sized polymethyl-methacrylate spheres, 0.3 μm in diameter and stabilized in suspension by a thin coating of another polymer. The organic suspensions were of high volume fractions, ranging from 0.39 to 0.53. After mechanical homogenization, each suspension was allowed to settle. In the volume fraction range 0.44--0.50, polycrystalline structures have bright opalescence; suspensions below 0.41 do not crystallize at all; whereas in the range 0.41--0.44, a stable liquid-plus-crystal two-phase state forms. But the really novel feature occurs in volume fractions above 0.50, where the viscous suspension remains in a 'glassy' form indefinitely: in effect, this is a kinetically metastable random dispersion which is too dense, and therefore too immobile, to order. Unlike two-dimensional bubble rafts, therefore,in three dimensions it is possible to create a glass from a population of monosized spheres. (Whether this is possible depends on the mobility of the units; be they macroparticles or atoms, pure metals cannot be prevented from crystallizing even at cooling rates in excess of 10^{12} $K\ s^{-1}$). Pusey and van Megen's method could be used to make a form of artificial opal if a way to stabilize the crystals against accidental randomization can be found.

Pusey and van Megen's suspensions differ radically from another kind which also show unexpected properties: Clark and Ackerson[10,11] demonstrated that very dilute (0.1 per cent) suspensions of highly charged polymer spheres can 'crystallize' into body-centred cubic crystals while remaining suspended, but randomize (i.e., melt) after vigorous shear deformation — an observation which could shed light on mechanical instability models of melting for ordinary atomic crystals.

The other paper in this issue, by Georges *et al.*. [2], exploits bubble rafts to simulate, in two dimensions, a microhard- ness indentation test. A conical indenter is simulated by a suitably shaped (poly)crystalline raft; the test object is a large monosized polycrystalline raft covered by an amorphous 'coating' consisting of a mix of two bubble sizes. The authors do not explain why they chose to test a glass-coated crystal-simulating sample, but in the light of some recent tests on real glassy wear-resistant coatings, it was a happy choice. The simulated coating behaviour is intimately affected by its crystalline substrate, which undergoes a largely elastic deformation. On unloading, this elastic distortion recovers and the form of the indentation recovers with it, so much so that the end-result is a small mound instead of an indentation.

This model system with its properties is of considerable interest in relation to recent work on the use of an ultra- microhardness tester, the nanoindenter, invented and named by W.C. Oliver[12,13]. This instrument is extremely sensitive, using measurements of the vertical displacement of the indenter (as small as 20 nm) to evaluate the residual indentation. (Direct measurement of such minute features would require scanning electron microscopy.) The instrument allows the increase in effective hardness for indents less than 100 nm deep to be confirmed and interpreted[12]. It was also found that considerable elastic recovery of the indentation takes place, and that this is essentially caused by a recovery of the 'hinterland' of the indentation, just as in the bubble simulation. But nanoindentations in real crystals do not invert and finish up as mounds.

Oliver and his collaborators have also studied the nanohardness of coated (ion-implanted) metals[13]. Of relevance to the present discussion is the behaviour of steel ion-implanted with both titanium (Ti) and carbon (C). If the concentrations of both are sufficient, the coating is very hard and wear-resistant. Such Ti/C implants are known to be amorphous, and the excellent mechanical properties of the layers are directly correlated with their vitreous nature. Georges[2] agree with these findings in the sense that their amorphous simulated coating proved impenetrable, being always pushed ahead of the moving indenter, but disagree in the sense that the yield stress they observed for the coating was lower than that for the crystalline substrate. Amorphous Ti/C coatings are stronger, harder and more wear-resistant than the underlying steel.

1. P.N. Pusey and W. van Megen, *Nature* **320**, 340 (1986).
2. J.M. Georges, G. Meille, J.L. Loubet and A.M. Tolen, *Nature* **320**, 342 (1986).
3. A.W. Simpson and P. Hodkinson, *Nature* **337**, 320 (1972).
4. A.S. Argon and H.Y. Kuo, *Mater. Sci. Engng.* **39**, 101 (1979).
5. P. Pieranski, *Contemp. Phys.* **24**, 25 (1983).
6. P.J. Darragh, A.J. Gaskin and J.V. Sanders, *Scient. Am.* **234** (4), 84 (1976).
7. B.W. Anderson, *Gem Testing* , 9th edn. (Butterworth, London,

1980).

8. M.D. Sacks and T.-Y. Tseng, *J. Am. Ceram. Soc.* **67**, 526 (1984).
9. P.D. Calvert, *Nature, News and Views* **317**, 201 (1985).
10. N.A. Clark, A.J. Hurd and B.J. Ackerson, *Nature* **281**, 57 (1979).
11. B.J. Ackerson and N.A. Clark, *Phys. Rev. Lett.* **46**, 123 (1981).
12. J.B. Pethica, R. Hutchings and W.C. Oliver, *Phil. Mag. A* **48**, 593 (1983).
13. J.B. Pethica, R. Hutchings and W.C. Oliver, *Nucl. Instrum. Meth..* **209/ 210**, 995 (1983).

9

The Supermodulus Syndrome

Nature, **324**, 108 (1986)

Synthetic compositionally modulated epitaxial semiconductor structures, or semiconductor superlattices, were first proposed by Esaki and Tsu in 1969. Because of their remarkable electronic and optical properties, these structures have been investigated intensively since then. The metallic equivalent of semiconductor superlattices, synthetic modulated metal films, began to excite interest in 1977, but until now studies on these films have been scientific rather than technological. Recent work has succeeded in using synthetic, bimetallic modulated films for the spectral analysis of X rays[1], the investigation of extremely low diffusivities of one metal in another[2,3] and the study of phase transformations[4].

In 1977, Yang *et al.* [5] prepared a series of multilayer films of the metal pairs Au–Ni and Cu–Pd. The elastic stiffness of the films was measured by a primitive method that allowed the film to bulge under gas pressure over a small hole; the bulge height is determined as a function of gas pressure. The stiffness thus measured is a biaxial modulus. For Au–Ni, the stiffness doubles for a critical thickness near 2 nm; for Cu–Pd, the effect is less pronounced. Recently, Baral *et al.* [6] used a more conventional, vibrating-reed approach to measure Young's modulus. The figures (next page) show their results[7] for Cu--Ni of constant 50/50 Cu/Ni ratio but varying periodicity and, for comparison, results with measurements of surface-wave velocity for 50/50 Cu–Nb films. The elastic anomaly near 2 nm thickness can evidently be a maximum or a minimum, although taken together, all published measurements[8] indicate that the anomaly is usually an enhancement and only rarely (to adopt the latest neologism) a dehancement. In Cu–Au films, however, there is no change. In some of the systems that show a modulus peak at a critical periodcity, there is also a peak of critical stress for plastic deformation.

There are two main rival theories that seek to interpret this 'supermodulus' effect. Clapp[9] proposed a purely electronic interpretation in which the compositional modulation creates additional Brillouin zones which can touch the Fermi surface and thereby stabilize the modulated structure. If this happens, then an elastic strain will tend to drag the zone and Fermi surface out of contact, which acts in the direction of structural destabilization and thereby enhances the stiffness. (The converse effect is also conceivable.) As various theorists have pointed out (for example, Falco *et al.* [10]), this hypothesis predicts several critical periodicities that should lead to anomalously enhanced (or dehanced) stiffness.

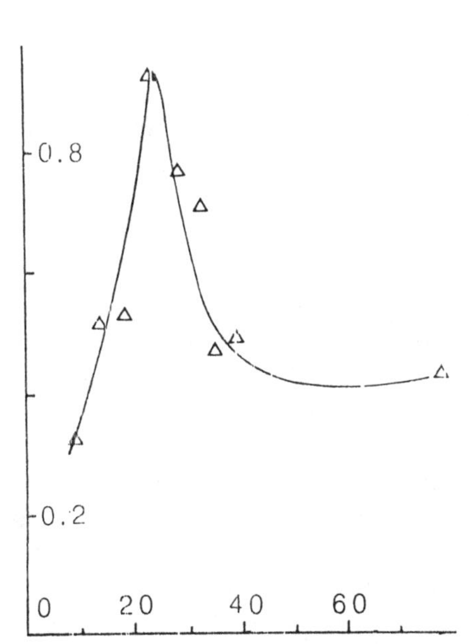

Left: variation of flexural modulus (in TPa, vertical scale) with wavelength (in Å, horizontal scale) for Cu-Ni. Below: surface-wave velocity (× 10^5 cm s^{-1}, vertical scale) of Nb-Cu superlattices as a function of bilayer thickness (in Å, horizontal scale).

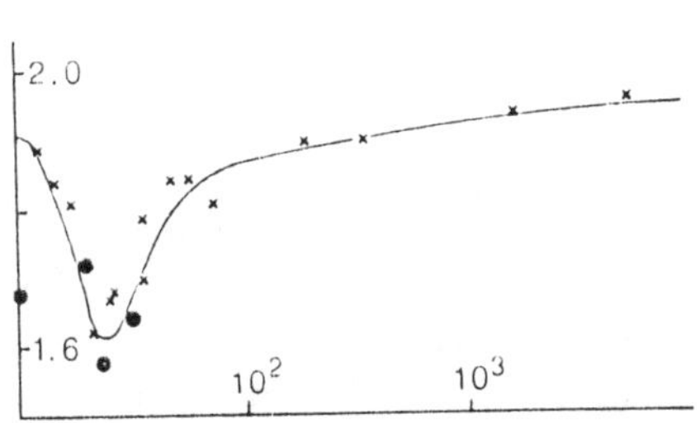

The alternative view[11] is that the super modulus effect is caused by coherency strains at the metal/metal interfaces, which are postulated to displace atoms so far from their normal sites that they are no longer within the hookean parts of their interaction potential. This approach can also interpret both enhancements and dehancements, but cannot explain the origin of the sharpness of the maximum so easily. For large periodicities, interface disloctions tend to form and release the coherency strains, but this still leaves unexplained the behaviour at periodicities of less than 2 nm. One interesting possibility is that the dehancement of stiffness in Cu-- Nb films might result from glass formation by interdiffusion, the so-called solid-state amorphization reaction[12]. Cu--Nb would be a reasonable candidate system for this process, though it has not been investigated with the amorphization reaction speciflcally in view. It is interesting that Lin and Spaepen[13] find that in Ni--Nb multilayer films, metallic glass is formed spontaneously during film deposition (that is, without anneal) for layer thicknesses smaller than 3 nm.

Very recent work using Cu–(Ni,Fe) multilayers[14] shows the existence of a minimum grain size (≈0.1 μm) at 2.6 nm periodicity. At this periodicity, the structure is seen to be defect-free; at smaller periodicities, partial interdiffusion reduces the sharpness of the interfaces and allows grains to coarsen; at periodicities greater than 2.6 nm, interfacial dislocations reduce the elastic coherency strains. These mechanisms can account for the modulus maximum at 2.6 nm. In other recent work[15], elastic moduli are calculated on the basis of the coherency strain model for Cu–Au and Cu–Ni multilayers as a function of periodicity distance, and roughly agree with experimental values.

The debate between the electronic and coherency strain hypotheses (see refs 8,11) remains wide open, but it does increasingly look as though the coherency strain model will turn out to be correct.

No practical use has been made yet of supermodulus enhancement, but the situation may be about to change. Stiffness is a major concern in the develop ment of improved aircraft alloys, and Al–Li and Al–Li–Be alloys, which are stiff and light, are now receiving concentrated attention. A very new approach is to make massive aluminium alloy pieces, large enough to machine wing spars, by sequen- tial evaporation[16] such alloys have substantially enhanced flow stress as well as stiffness, but there has been no report of a supermodulus effect at critical periodicities. In any case, the periodicities used were 30-40 times larger than in the thin-film work, where peaks in flow stress have been observed at ≈ 2 nm periodicities[15]. It remains to be seen whether alloy pieces of useful size can be made with perodicities around 2 nm, and whether an aluminium-based system will show the supermodulus effect. Yahalom[17] has just reported a very promising new approach to making nanometre metallic multilayers by pulsed electrodeposition, applied initially to the Cu–Ni system. He specifically suggests applying his method to the production of springs.

1. J.H. Underwood and T.W. Barbee, *Nature* **194** , 492 (1981).
2. A.L. Greer, *Scripta metall.* **20**, 457 (1986).
3. R.C. Cammarata and A.L. Greer, *J. Non-Cryst. Solids* **61/ 62**, 889 (1984).

4. T. Tsakalakos, *Scripta metall.* **20,** 471 (1986).
5. W.M.C. Yang, T. Tsakalakos and J.E. Hilliard, *J. appl. Phys.* **48,** 876 (1977).
6. D. Baral, J.B. Ketterson and J.E. Hilliard, *J. appl. Phys.* **57,** 1076 (1985).
7. A. Kueny *et al.*, *Phys. Rev. Lett.* **48** , 166 (1982).
8. R.C. Cammarata, *Scripta metall.* **20,** 479 (1986).
9. P.C. Clapp, in*Modulated Structure Materials* (ed. T. Tsakalakos) 465 (Nijhoff, Dordrecht, 1984).
10. C.M. Falco and I.K. Schuller, in *Synthetic Modultaed Structures* (eds. L.L. Chang and B.C. Giessen) 339 (Academic, London, 1985).
11. T. Tsakalakos and A.F. Jankowski, *Ann. Rev. Mat. Sci.* **16,** 293 (1986).
12. R.B. Schwarz and W.L. Johnson, *Phys. Rev. Lett.* **51,** 415 (1985).
13. C.J. Lin and F. Spaepen, *Acta metall.* **34,** 1367 (1986).
14. A.F. Jankowski and J.F. Shewbridge, *Mat. Lett.* **4,** 313 (1986).
15. M. Imafuku, Y. Sasajima, R. Yamamoto and M. Doyama, *J. Phys. F* **16,** 823 (1986).
16. R.L. Bickerdike *et al.* , *Int. J. Rapid Solidification* **1,** 305 (1986).
17. J. Yahalom, in *Materials and Prtocessing Report* **1,** no. 7 (October 1986).

10

Strategies to defeat Brittleness

Nature, **332**, 112 (1988)

The highest attainable strength is found in materials composed of light atoms linked into a network by strong ionic or covalent bonds — in other words, ceramics. These also have particularly high melting-points and so would be apt for engineering use at both ambient and high temperatures were it not for their universal brittleness, which is a synonym for the inability to deform plastically and thus to absorb mechanical or thermal shocks. No ultrabrittle material could ever be trusted, for instance, in an aero engine. Two recent papers[1,2] present radically new ways of overcoming brittleness in hard materials. Gleiter *et al.* [1] show that reducing the grain size of ceramics can allow them to flow plastically. And on page 139 of this issue[2], Brookes *et al.* show that even diamond, if pressurized in the right way, can deform without fracturing.

During the past 15 years particularly, attempts have been made to overcome the handicap of brittleness by improvements in processing and composition[3]. Macro-defect–free cement[4], chemically modified silicon nitrides[5] and transformation-toughened zirconia[6,7] represent successful improvements based on these two strategies. The last of these was originally proposed[8] with the provocative title 'Ceramic steel?', because doped zirconia, like high-tensile steel, depends on a partial phase transformation to temper strength with shock resistance: an advancing crack triggers a local transformation which hinders further propagation of the crack.

None of these improvements has gone far enough. Jack[5] described with feeling the technological, economic and organizational problems that dimmed the promise of silicon nitride, which "has now been the ceramic of the decade for three decades". The ceramic engine, outside Japan at least, still seems a long way from practical realization. There is scope for radically new approaches.

One such approach, recently reported by Gleiter and co-workers in *Nature*[1], depends on the creation of what is virtually a new state of matter, intermediate between a polycrystal and a glass — the 'nanocrystal". The structure of a nano-crystalline metal, made by evaporation or sputtering onto a cold substrate followed by compaction of the fine powder so produced, is shown schematically in the figure. The crystal grains are so small that almost half the atoms in the metal occupy intergranular sites; examination by X-ray diffraction of nanocrystalline gold shows[9] that the intergranular material is gas-like rather than glass-like — there is no short-range order. Mössbauer spectroscopy of nanocrystalline iron over a range of temperatures shows[10] that the intergranular matter has a lower Debye temperature (or elastic constant), and thus has a lower cohesion, than do the grains.

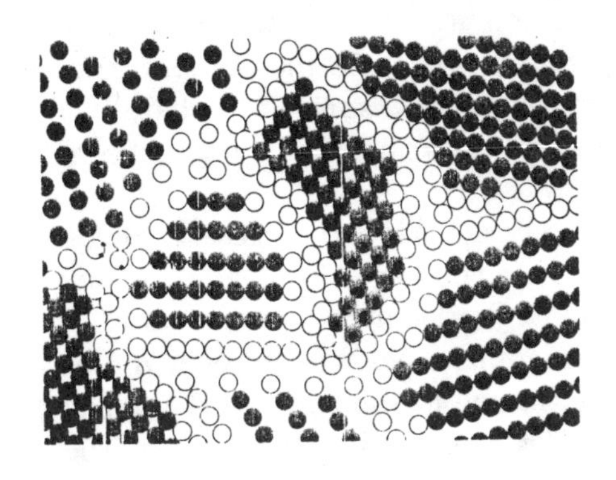

Structure of a nanocrystalline metal. Solid circles, atoms arranged in crystalline structures; open circles, loosely packed intergranular atoms that give the material its high self-diffusivity and plasticity. Under pressure, atoms on the compressed side of a grain rapidly self-diffuse through the intergranular circuit to the stretched side. (Courtesy of H. Gleiter.)

The solubility of hydrogen as a function of its concentration is a measure of the prevalence of microcavities, which are hydrogen traps; such solubility measurements on nanocrystalline palladium show[11] that the intergranular 'phase' is indeed full of cavity-traps. This in turn implies, as has been observed[11], that solute diffusion of small atoms along the intergranular short-circuits is extremely concentration-dependent and can be higher or lower than in the crystal. Self-diffusion, however, is not affected by traps and is spectacularly enhanced. Self-diffusion at 20-120°C in nanocrystalline copper is enhanced by no less than 19 orders of magnitude compared with an ordinary polycrystal[12].

Gleiter and co-workers[1] observed easy plasticity, at 80-180°C, of nanocrystalline TiO_2 (made by oxidation of nanocrystalline titanium) and CaF_2 with grain diameters of about 10 nm — smaller by a factor of about 1 000 than normal grains. This can be interpreted in terms of diffusion creep. In this process, the cations and anions diffuse from the compressed to the stretched side of each grain, round an intergranular circuit, thereby changing the shape of the grain. The authors estimate that the small grain size enhances the strain rate by a factor of 10^9s, and the large self-diffusivity by a further factor of 1 000. In effect, these materials behave superplastically[13], although the superplastic strain rates that are shown to be feasible are very much higher than those normally encountered in the industrial superplastic forming of aluminium alloys.

Of course, a superplastic ceramic will be weak in a critical range of temperatures: but this can in principle be counteracted by heating the ceramic to make the grains grow slightly: the larger the grains, the higher the critical temperature range. As the authors point out, the core and surface of a ceramic block can be treated (perhaps by microwave heating) to have different grain sizes and plastic properties. The possibilities are intriguing.

Gleiter and colleagues found it necessary to deform their ceramics by compressing them in contact with a platen made of a soft metal, aluminium, backed by a harder metal; presumably, contact with a hard platen causes intragranular cracking before the diffusion creep can begin. This soft-platen technique is similar to that used in the very different study of Brookes *et al.* reported in this issue[2]. These authors achieve plastic deformation in a monocrystal of diamond, the hardest and most brittle of solids. Following earlier work with MgO and TiC, they reason that a conical indenter pressed on a smooth diamond surface under constant load will — so long as it is slightly softer than the diamond — slowly be blunted until the distributed contact stress has fallen sufficiently for deformation to cease. If the hardness of the indenter is chosen correctly, then the stress in the contact zone is too low to crack the diamond but high enough to deform it plastically. For this to be feasible, the temperature must be very high. An indenter of cubic boron nitride (second only to diamond in hardness) at 1 000°C neatly does the trick. The diamond deforms by clearly visible multiple dislocation glide. This temperature is 500°C lower than the hitherto accepted transition from brittleness to slight ductility, as determined by three-point bending[14].

It would be interesting to know how the grain size of pol7crystalline diamond affects its plastic deformability; indeed, it is conceivable that a sufficiently fine grain size could permit a measure of diffusion creep, not linked to the motion of dislocations. In view of the extensive recent work on the formation of microcrystalline diamond surface coatings by vapour deposition[15], such an investigation could be of great practical value.

1. J. Karch, R. Birringer and H.Gleiter, *Nature* **330**, 556 (1987).
2. C.A. Brookes, V.R. Howes and A.R. Parry, *Nature* **332**, 139 (1988).
3. R. Brook, in *Encyclopedia of Materials Science and Engineering* (ed. M.B. Bever) 98 (Pergamon and MIT, 1986).
4. J.D. Birchall, *Phil. Trans. R. Soc. (Lond.) A* **310**, 139 (1983).
5. K.H. Jack, in *High-Technology Ceramics, Past, Present and Future; Ceramics and Civilization* Vol. III (ed. W.D. Kingery) 259 (Am. Ceram. Soc., Westerville, 1986).
6. N. Clausen adn A.H. Heuer, in *Encyclopedia of Materials Science and Engineering* (ed. M.B. Bever) 5129 (Pergamon and MIT).
7. F.F. Lange, *J. Mater. Sci.* **17**, 225 (1982).
8. R.C. Garvie, R.H. Hannink and R.T. Pascoe, *Nature* **258**, 703 (1975).
9. X. Zhu, R. Birringer and H. Gleiter, *Phys. Rev. B* **35**, 9085 (1987).
10. U. Herr, J. Jing, R. Birringer and H. Gleiter, *J. app. Phys.* **50**, 472 (1987).
11. T. Mütschele and R. Kirchheim, *Scripta metall.* **21**, 135 (1987).
12. J. Horvath, R. Birringer and H. Gleiter, *Solid-State Commun.* **62**, 319 (1987).
13. R.W. Cahn and P. Hazzledine, in *Encyclopedia of Materials Science and Engineering* (M.B. Bever, ed.) 4686 (Pergamon and MIT, 1986).
14. T. Evans and R.K. Wild, *Phil. Mag.* **12**, 479 (1965).
15. R. Messier, K.E. Spear, A.R. Badzian and R. Roy, *J. Metals, N.Y.* **39** (9), 8 (1987).

11

Keeping Dispersions Fine

Nature, **336**, 19 (1988)

High-temperature alloys, especially those used in aero engines, have to serve at temperatures well above half their absolute melting point. Extremely fine dispersions of insoluble intermetallic phases that impede dislocation motion are essential to prevent failure, the only alternative being strengthening by ceramic fibres, a strategy that is still being developed. The fine dispersion must resist coarsening for a particular matrix/dispersoid combination to be useful. Ostwald ripening, the growth of larger particles at the expense of smaller ones by a process of solute diffusion through the matrix, is not important if the particles are very insoluble in the matrix. But an alternative process, the dragging of particles attached to the migrating boundaries of growing grains, can still force coarsening: the particles which are being dragged sweep up stationary particles from the body of a grain as the boundary moves past them. This process has not been analysed as much as it deserves, and hence three recent papers on the subject by Russell and Froes[1,2]and by Hartland *et al.* [3] deserve a special welcome.

Russell and Froes[1], who analyse the thermodynamic factors that favour the creation of a fine-particle dispersion in the first place, discuss the probable rate of Ostwald ripening. They also analyse the limiting rate at which a particle can keep up with a moving grain boundary; this depends both on the particle diameter and on its solubility in the matrix. The authors conclude that a drift rate of 1 $\mu m\ h^{-1}$ can be maintained by particles of 1, 10 and 100 nm radius for low (10^{-4} atom %), intermediate (0.1 atom %) and high (10 atom %) solubilities, respectively. This is fast enough to lead to a potentially dangerous degree of sweeping, at least for the larger particles.

The analysis is dealt with in greater detail in Russell and Froes's second paper[2]. The calculation is done on the assumption that the particles migrate by the process of diffusion along the particle/matrix interface, presumed to be faster than diffusion through the bulk of the particle. There is still uncertainty, dating back to a classic paper by Ashby and Centamore[4], about whether crystalline particles can drift at anything like the rate of amorphous particles (such as silica), because crystalline particles tend to have a more ordered interface with the matrix, associated with slower diffusivities in the interface.

Russell and Froes compare experimental observations of various dispersoid systems with their calculations, with special reterence to the important case of Ti_3Al strengthened with dispersions of Er_2O_3 or Ce_2S_3. It turns out that the greater resistance of the latter to coarsening can be explained in terms of the characteristics of the two dispersoids.

Hartland *et al.* [3] study the special case of a population of empty or gas-filled pores drifting with a migrating boundary. This theme is older than the problem of particle drift: inclusion drift in a crystalline material was first observed for helium bubbles in copper[5], and this led to the first theoretical treatment of bubble drift in a temperature gradient[6] (without attachment to grain boundaries). The possible role of grain boundaries in dragging pores during the process of powder sintering and thereby hastening the elimination of porosity was pointed out several years ago[7], but evidence of such dragging is very scant and uncertain.

Hartland *et a1.* [3] are the first to devise a theory of the dragging of pores by boundaries. Their special interest is the behaviour of fission-gas pores in uranium dioxide nuclear fuel. The authors analyse in rigorous detail the behaviour of various grain and pore geometries, including a range of porosity fractions and sizes, and for each situation calculate the retardation factor imposed by the presence of pores, compared to an identical grain geometry without pores. One factor they take into account (this has never been considered before) is the influence of misorientation across a grain boundary; this affects the curvature of the interface of a trapped pore and thereby influences surface diffusion along that interface. It also turns out that pores retard grain boundary mobility more on grain edges than when at grain corners.

One variable not considered in this work is the effect of impurities dissolved in the matrix segregating to the pore/matrix interface; it has been shown experimentaly, in metallic systems[8], that such segregation can enhance surface diffusivity by factors of up to 10^4. In practical terms, it would be interesting to know whether segrcgation of this kind is possible to the interface between dispersoid particles and the matrix. If so, it is conceivable that dispersoid mobility and thus dispersoid coarsening through drift of particles attached to grain boundairies could be considerably enhanced in the presence of particular impurities.

1. K.C. Russell and F.H. Froes, *J. Metals* **40**, 29 (1988).
2. K.C. Russell and F.H. Froes, *Scripta metall.* **22**, 495 (1988).
3. P. Hartland, A.G. Crocker and M.O. Tucker, *J. Nucl. Mater.* **152**, 310 (1988).
4. M.F. Ashby and R.M.A.Centamore, *Acta metall.* **15**, 1081 (1968).
5. R.S. Barnes and D.J. Mazey, *Proc. R. Soc. (Lond.) A* **275**, 47 (1963).
6. P.G. Shewmon, *Trans. Amer. Inst. Min. Metall. Engrs.* **230**, 1134 (1964).
7. A.H. Heuer, *J. Am. Ceram. Soc.* **62**, 317 (1979).
8. G.E. Rhead, *Surface Sci.* **47**, 207 (1975).

12

Fractal Dimension and Fracture

Nature, **338**, 201 (1989)

How rough is a fracture surface? This question is in the domain of the craft of fractography, which seeks to elucidate the mechanisms of fracture by studying the topography of a broken object. Somewhat akin to phrenology, some sceptics might say. Nevertheless, in two new studies, Mecholsky, with Mackin[1] and with Passoja and Feinberg-Ringel[2], shows that it may be possible to scale the fracture toughness of fairly brittle materials by just such an exercise in fractographic analysis, exploiting the concept of fractals.

Fractals were invented — or at any rate clearly formalized — by Mandelbrot[3,4]. The term refers to any boundary or surface which remains self-similar as the scale of examination is magnified. Figure 1 gives an example of such a boundary. For this shape, a 'Koch island', drawn to be always of constant area, the length of the perimeter is increased by a factor of 1.5 from one drawing to the next if the length of the ruler unit used to measure it is decreased by half, and the image resolution grows by a factor of four. If the logarithm of the total perimeter length is plotted against the logarithm of the length of the ruler unit (the resolution), a straight line or 'Richardson plot' is obtained, the slope of which was treated by Mandelbrot as a fractional, *fractal* , dimension: for a boundary as in Fig. 1, this dimension is between 1 and 2 — in this case, 1.5.

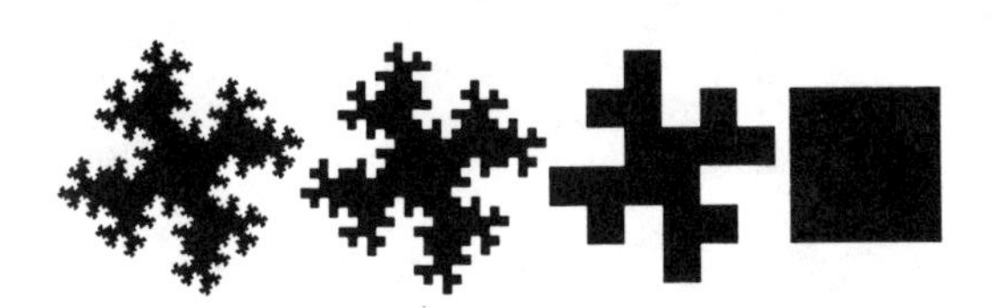

Fig. 1. A Koch island with a perimeter of fractal dimension 1.5 (ref. 5)

Similarly, a rough surface which is self-similar has a dimension between 2 and 3. To obtain this fractal dimension, it is usual either to make serial sections perpendicular to the surface ('vertical' sections) and examine the shape of the section boundary, or to section parallel to the surface and examine the perimeters of the 'islands' and 'lakes' which are in effect contours of such 'horizontal' sections.

The original concept of fractals was restricted to self-similar outlines only, but more recently the concept has been extended to more ordinary boundaries and surfaces of varying roughness; thus, Fig. 2 shows four outlines generated by Russ[5] by a rather involved computer algorithm, with four different fractal dimensions. In such cases, the corresponding Richardson plot has a straight central part with curved regions at either end. (Such shapes should really be called 'pseudo-fractals', but that term does not appear to have been used.) Stereologists still disagree about the appositeness of the horizonal section approach[6], but in a recent study of fracture surfaces of ceramics, Mecholsky and Passoja[7] found that the horizontal and vertical methods gave closely similar fractal dimensions.

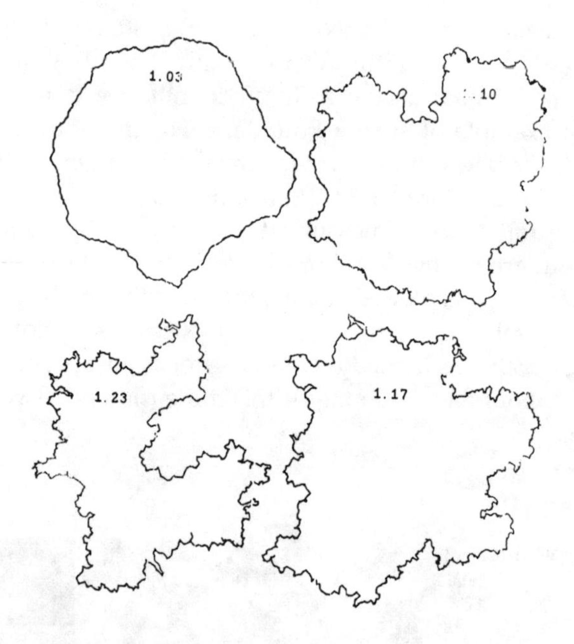

Fig. 2. A random fractal outline (ref. 5).

The studies of Mecholsky and colleagues[1,2,7] on the relation between fractals and fracture were stimulated by two apparently independent initiatives. One was a fractographic study by Mandelbrot *et al.* [8] of broken specimens of maraging steel. The impact toughness of 300-grade maraging steel in various states of heat-treatment increases steadily as the fractal dimension of the fracture surface decreases towards 2 (i.e., it becomes steadily smoother)[8]. It is apposite that Mandelbrot should have been involved in this early study because, as he says, he coined the term fractal "in explicit cognizance of the fact that the irregularities found in fractal sets are often strikingly reminiscent of fracture surfaces in metals (though not, for example, in glass)". Although he does not say so, one should not lose sight of the fact that the word 'fraction' has the same Latin root as 'fracture', and fractals have fractional dimensions.

The second study was by Beauchamp and Purdy[9], who examined the changes in fracture toughness and fracture morphology of chert (flint) as it is heat-treated at progressively higher temperatures. It turns out that chert heated to 500°C is considerably less tough than untreated material; the heated chert also has a much smoother surface. (This implies that heated chert is easier to chip and shape into tools, and indeed there is archaeological evidence that chert was commonly heated in prehistoric times.) Heating coarsens the exceedingly fine grain of the chert and also apparently enhances the strength of the intergranular bond: un-heated chert fractures along the boundaries of the fine grains, heated chert breaks transgranularly. But the unheated material is nevertheless tougher because fracture is constrained to follow a very tortuous path.

Beauchamp and Purdy did not explicitly invoke the fractal concept in their study, but in the new work[1,2] by Mecholsky and co-workers, their original samples were subjected to a proper fractal analysis, using both horizonal and vertical methods of analysis. It is quite clear that the fracture toughness, K_c , of chert increases as the fractal dimension increases — as the fracture surface becomes rougher. This is the exact converse of what was found for maraging steel. (The relationship for chert is in the form of a linear log--log plot of K_c against fractal dimension.)

The authors also indicate, by means of a figure relating the fractal dimension of the fracture surface to fracture toughness for several ceramics (oxides, silicates, chal cogenides), that this form of relationship is universal for such highly brittle materials, though the results bunch into several distinct material groups rather than falling along a single master line. This research is itself a continuation of earlier work[7].

We thus have the intriguing fact that the relationship between fracture roughness (fractal dimension) and fracture toughness is diametrically opposite for ductile fracture (maraging steel) and for brittle fracture (chert and polycrystalline ceramics). For chert, where fracture is governed by the spreading of cracks, the toughening influence of factors making for rough fractures has been qualitatively explained. For ductile fracture, which can take place by the progressive creation, growth and merging of minute voids, the relationship is clearly different. Mandelbrot *et al.* [8] seek to interpret their findings in terms of percolation concepts applied to the filaments remaining intact between voids, but it must be said that their argu ment is not easy to follow. It is to be hoped that in future work, Mecholsky and colleagues will address the piquant distinctions between ductile and brittle fracture which these studies have uncovered.

1. J.J. Mecholsky and T.J. Mackin, *J. Mater. Sci. Lett.* **7,** 1145 (1988).
2. J.J. Mecholsky, D.E. Passoja and K.S. Feinberg-Ringel, *J. Am. Ceram. Soc.* **72,** 60 (1989).
3. B.B. Mandelbrot, *The Fractal Geometry of Nature* (Freeman, New York, 1982).
4. J. Gleick, *Chaos: Making a New Science* , 94-118 (Heinemann, London, 1988).
5. J.C. Russ, in *Encyclopedia of Materials Science and Engineering , Suppl. Vol. 1* (ed. R.W. Cahn) 169-175 (Pergamon, Oxford, 1988).
6. E.E. Underwood and K. Banerji, *Mater. Sci. Engng.* **80,** 1 (1986).
7. J.J. Mecholsky and D.E. Passoja, in *Fractal Aspects of Materials* 117-119 (Mater. Res. Soc., Pittsburgh, 1986).
8. B.B. Mandelbrot, D.E. Passoja and A.J. Paullay, *Nature* **308,** 721 (1984).
9. E.K. Beauchamp and B.A. Purdy, *J. Mater. Sci.* **21,** 1963 (1986).

13

Nanostructured Materials

Nature, **348**, 389 (1990)

When a new field of research arrives upon the public scene, the first thing apt to excite dissension is either the naming of the key phenomenon, or else its very existence. Following the recent claims concerning fusion of deuterium dissolved in metallic palladium, the term 'cold fusion' was happily accepted by all, but the phenomenon was not. On the other hand, although the theme of a recent conference entitled *Materials with Ultrafine Microstructures* [1] was agreed as being sound enough by everyone present, the proper nomenclature was not agreed.

The subject of the conference, the first full-scale meeting on this theme, was the synthesis, characterization and properties of polycrystalline materials, the constituent grains of which are only a few nanometres across (typically, $2\text{--}25 \times 10^{-9}$ m). The originator of this field of research, H. Gleiter (University of Saarbrücken) coined the term 'nanocrystalline materials'. Because such materials are usually made by consolidating minute clusters of atoms, the term 'cluster-assembled materials' was the next to be proposed[2]. As the title of the recent conference indicates, others like the sound of 'materials with ultrafine microstructure', while a further, slightly eccentric group proposes the term 'finite systems' to embrace clusters and cluster assemblies and quasicrystals for good measure. But 'nanostructured materials', the title of a projected international meeting in 1992, may be the finally accepted description.

Clusters

The main emphasis of the meeting was on synthesis, processing and applications. Until recently, Gleiter's original method of making the precursor clusters, by evaporating a metal or ceramic into a noble gas at fairly high pressure, restricted quantities to fractions of a gram. Kratschmer and colleagues recently showed[3] how buckminsterfullerenes such as the spherical cluster C_{60} could be made in gram quantities by heating a carbon rod electrically and condensing the resulting smoke. (C_{60} has captured popular imagination to such a degree that a cartoon recently had the hero thrashing about helplessly in a vat of low-friction C_{60} fullerene.)

Clearly, the urge to scale up is now widespread. For example, the straight evaporation method has been developed to the point where useful nanostructured metallic films can be deposited on a silicon substrate and aerosol clusters can be directed to form conducting strips more than 50 μm wide at specific locations on microcircuits (M. Oda, Vacuum Metallurgical Company, Japan). These films and wires have good adhesion and conduct well. Oda claimed that his method had distinct advantages over conventional methods of forming 'wiring' on microcircuits. The size of circuit elements made in this way is being steadily reduced, and (this being Japan) the equipment needed is already for sale. M. Uda (Nisshin Steel Company, Japan) showed that the rate of generating metallic clusters into hydrogen gas, in preference to argon, could be enhanced by several orders of magnitude by first activating the hydrogen electrically. The lone Soviet participant, L. Trusov (Research Industrial Enterprise Ultram, Moscow), though he was coy about his methods of synthesis, showed favoured participants a large metallic nanofilter sheet made from evaporated and consolidated clusters.

Combination

Probably the most impressive account of synthesis and application combined came from B. Kear's group (L. McCandlish, Rutgers University). They set out to improve the 50-year-old 'hard metal' which consists of rather coarse, hard and brittle tungsten carbide crystals dispersed in a soft, ductile matrix of cobalt metal — a cermet. They guessed that if they could make such a material with nano-sized crystallites, its mechanical properties might be enhanced. Enlisting the aid of a chemist, the group combined *tris* - (ethylenediamine) cobalt tungstate with ammonium metatungstate and cobalt chloride, carburized the product first with CO and then with a CO/CO_2 mixture (a sequence which permitted them to combine rigorous control of C activity with rapid production), and then consolidated the mixed product to form nanostructured hard metal. The new material is up to 40 per cent harder than conventional hard metal for a given cobalt content and the friction and wear rates are much lower; high hardness and low wear rate is the required combination for machining tools. An industrial-scale plant to make the new material has now been designed round the use of a fluid-bed reactor.

Also from Rutgers University, S. Danforth analysed in detail the benefits flowing from the use of nanostructured silicon nitride (one of the front-running candidates for ceramic automotive engine components). Essentially, such a fine powder can be sintered at a much lower temperature than conventional powders and this permits the product's porosity to be depressed to low levels and its consequent properties, especially at very high temperatures, to be improved. The ready sinterability of nanostructured powders formed the theme of several other papers and is clearly an important research topic: it is important that grain growth is often surprisingly sluggish so that nanosized powders can be sintered without coarsening much. The reasons for the slow growth constitute another current research theme.

Strength

The strength of nanocrystalline Cu and Pd seems hardly to depend on grain size d : the familiar Hall--Petch $d^{-1/2}$ dependence was generally not observed (J. Weertman, Northwestern University). A particularly novel study (V. Provenzano, Naval Research Laboratory, Washington, D.C.) building on much earlier studies by Marcus[4] (experimental) and Louat[5] (theoretical) shows that materials of quite exceptional strength — they are in fact termed 'superstrong' — can be obtained by embedding a high volume fraction of nanosized particles in a ductile matrix or alloy; in one version, the matrix is glassy. Apart from the ease of sintering nanosized powders, another advantage of consolidated nanostructured ceramics is that they can be superplastically deformed, as Gleiter first showed some years ago. This process, reported by several contributors and recently described in *Nature* [6], depends on the diffusion of lattice vacancies through or round small grains, effectively transporting matter to respond to applied stress.

Not only people take an interest in nanostructures: nature does also. There are fascinating lessons to be learned from the fine structure of seashells, abalone shells (*haliotis rubescens*) in particular (I. Aksay, Seattle University). One layer of such a shell consists of 250-nm laminations of aragonite (a polymorph of $CaCO_3$) separated by 20-nm layers of an organic compound. This structure generates both high strength and high fracture toughness. Years ago, useful lessons were learnt from the laminated fibre-reinforced structure of the wing-cases of rose chafer beetles; now[7], a B_4C/Al laminar composite has been made, following the abalone model, and proved to have outstanding properties (Fig. 1).

By no means was the conference limited to applications. Fundamental scientific issues were also much aired: for example, the nature of the intercrystalline boundary regions in nanostructured materials. Whereas high-resolution electron microscopy (G. Thomas and R. Siegel, Argonne National Laboratory) and Raman spectroscopy of

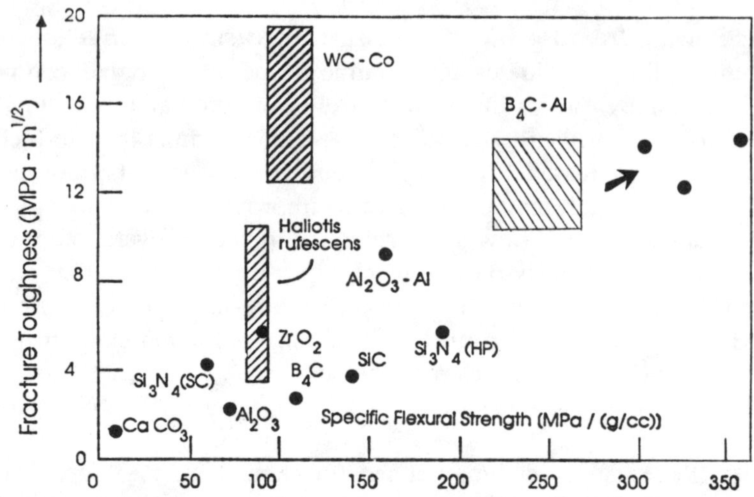

Fig. 1 Fracture toughness versus fracture strength (normalized with respect to density) of the nacre section of abalone shell compared with several high-technology ceramic and cermet materials. (After Sarakaya *et al.*[7].)

---

nanocrystalline CO_2 [8] indicate a normal, atomically narrow grain-boundary structure, Mössbauer studies (Gleiter; B. Fultz, California Institute of Technology), electrochemical measuremcnts (R. Kirchheim, Max-Planck-Institut für Metallforschung, Stuttgart, Germany), density and diffusion measurements (Gleiter) and calorimetry indicate that boundaries are more disordered and have a lower local density than conventional boundaries. The curiously sluggish grain growth already cited may be linked with this. Fultz and coworkers' Mössbauer data for a nanostructured Cr--Fe alloy can be accurately analysed to yield an estimated boundary width of about 1 nm; after grain growth, the boundary signal of course diminishes. The issue is still very much open; it may be that the structure depends sensitively on the nature of the misorientation and varies from boundary to boundary.

Several excellent papers were devoted to the characteristic clusters themselves. S. Riley (Argonne National Laboratory) showed how precise measurements of NH_3 uptake could be used to identify the exact geometrical form of size-separated metal clusters, Ni and Co in particular, because NH_3 attaches itself only to atoms standing proud of the surface, for instance at icosahedral vertices. X. Miao and P. Marquis (Birmingham University) showed how pH control and the combination of two differently sized populations (of alumina and silica) allowed them to create stable combined colloid particles of the two compounds which then sinter more easily.

Magnetic studies of both elementary and multicompound

nanoclusters, as well as metallic multilayers, reveal curious combinations of ferromagnetic, antiferromagnetic and superparamagnetic behaviour, as well as anomalous easy magnetization directions. In this connection, a highlight of the conference was H. Warlimont's report on recent developments by G. Herzer in his laboratory (Vacuumschmelze, Hanau, Germany) of an innovation first reported by Yoshizawa *et al.* [9]. Certain metallic glasses containing Fe, Si and B have excellent soft magnetic properties and are currently used as transformer laminations. Yoshizawa and co-workers found that if Nb and (crucially) Cu are added to this glass and it is then crystallized in the right way, a giant magnetic permeability can be attained, comparable with that of permalloy.

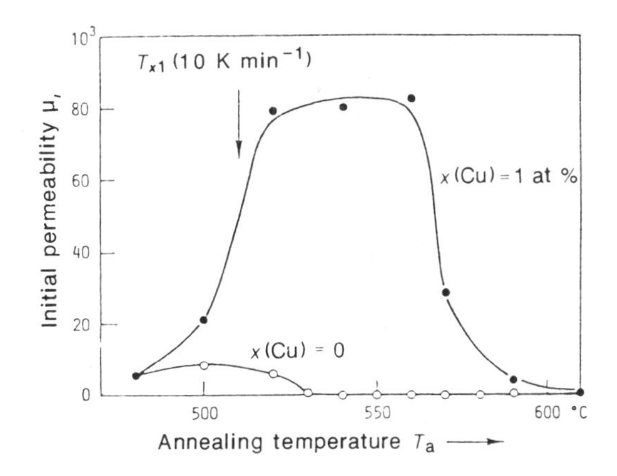

Fig. 2 Initial permeability. μ, of $Fe_{74.5-x}$ $Nb_3Si_{13.5}B_9$ alloys, initially glassy with x =0 and x = 1). as a function of annealing temperature, T, for a constant 1-h annealing time. Y_{x1} is the first crystallization temperature. (After G. Herzer and Warlimont.)

Herzer and Warlimont's results (Fig. 2) show that 1 atomic per cent of copper has a spectacular effect. It turns out that the annealed glass (provided it is not heated at too high a temperature) consists of nanocrystalline, ferromagnetic Fe_3Si phase separated by films of residual glass, rich in B and Cu, which is either nonmagnetic or has weaker magnetism. The residual glass crystallizes above 600°C. Herzer and Warlimont analysed the magnetic properties of such a structure.

Nanocrystalline ferromagnetic particles are single-domain (Gleiter in a recent review[10] cites very new evidence to this effect), and the ferromagnetic exchange interaction then overtrumps the effects of magnetocrystalline anisotropy in the nanograins and constrains the magnetization vector to be parallel to that in a neighbouring grain, irrespective of easy directions of magnetization. At a critical grain size, this results in very low coercivity and high permeability, as observed: here the critical dimension is around 35 nm. This Cu-bearing alloy has also a very low magnetostriction and the hysteresis loop shape can be shaped by magnetic annealing. This kind of nanostructured magnetic dispersion in a residual glass, by analogy with the 'superstrong' dispersions cited above, might perhaps be termed a 'super-magnet' — nanostructural researchers are predisposed to superlatives.

1. *Acta Metallurgica* conference on *Materials with Ultrafine Microstrctures* , Atlantic City, N.J., 1-5 October 1990. (Proceedings published in the new journal *Nanostructured Materials* , Pergamon Press, 1992).
2. R.P. Andres *et al., J. Mater. Research* **4**, 704 (1989).
3. W. Krätschmer, L.D. Lamb, K. Fostiropoulos and D.R. Huffman, *Nature* **347**, 354 (1990).
4. M.A. Marcus, *Acta metall.* **27**, 893 (1979).
5. N.P. Louat, *Acta metall.* **33**, 59 (1985).
6. F. Wakai *et al.* , *Nature* **344**, 421 (1990).
7. M. Sarikaya, K.E. Gunnison, M. Yasrebi and I.A. Aksay, *Mat. Res. Soc. Symp. Proc.* **174**, 109 (1990).
8. C.A. Melendres *et al.* , *J. Mater. Research* **4**, 1246 (1990).
9. Y, Yoshizawa, S. Oguma and K. Yamauchi, *J. Appl. Phys.* **64**, 6044 (1988).
10. H. Gleiter, *Progr. Mater. Sci.* **33**, 1 (1990).

Epitaxy

14

When is Epitaxy Possible?

Nature, **223**, 1206 (1969)

Epitaxial deposition — the growth of a film of a crystalline species in parallel orientation on a monocrystalline substrate of another species — is a well established technique which has become of major technological importance with the growth of the solid state electronics industry. Attempts to interpret the phenomenon, and in particular to define necessary and sufficient conditions for it, go back to the pioneering work of Frank and van der Merwe twenty years ago, and have concentrated on the lattice parameter misfit between film and substrate and its implications. More recent experimental work has established many cases of a distinct phenomenon, pseudomorphous deposition of thin films on monocrystalline substrates so as to mimic the substrate structure. Thus, iron deposited on a copper substrate will, in appropriate conditions, deposit epitaxially as a metastable face-centred cubic film; this is pseudomorphous, for it mimics the copper structure whereas iron in equilibrium at the deposition temperature is body-centred cubic.

W. A. Jesser, who was responsible for several experimental studies of thin-film pseudomorphism, has now worked out a detailed theory of the phenomenon[1]. Such a theory has to interpret not only the limiting conditions which permit pseudomorphism at all but also has to account for the limiting thickness beyond which an epitaxial film is apt to return to the stable bulk structure. The theory works up through a series of approximations to an analysis of a pseudomorphous epitaxial film on a substrate which contains a small hemispherical embryo of the thermodynamically stable form, in contact with the substrate along its flat surface and (at least for a small embryo radius) coherent with the substrate. The problem is to calculate, first, for what limiting pseudomorphous film thickness it is energetically favourable for the film to be coherent with the substrate. (For greater thickness, interface dislocations begin to appear.) Second, it is necessary to decide what is the critical growth radius of the hemispherical embryo of the normal structure, taking into account the change in the elastic strain energy of the film that results from the growth of the embryo. (If I have correctly understood the theory, this thickness would have to be less than the limiting thickness just mentioned.) It is the detailed calculation of this elastic energy component in the energy balance which is the chief contribution of this paper.

The theoretical equations are applied to Jesser's own observations on pseudomorphous iron films on copper. (He does not quote the film thickness and this parameter seems to receive rather cavalier treatment in the theory.) At room temperature, the pseudomorphous iron films are partly transformed into spherical regions of body-centred cubic iron: the smallest of these regions are 25 Å in radius, and this should thus match the calculated critical radius for body-centred cubic embryos to turn into viable nuclei. The value obtained from the theory is 30 Å, which represents very good agreement.

The next step would now be to establish a convincing theoretical criterion for the formation of an epitaxial deposit at all — whether normal or pseudomorphous. The naive idea of twenty years ago, following on Frank and van der Merwe's original ideas, that a misfit of about 14 per cent, with identity of crystal structure, is a neccssary and sufficient condition for normal epitaxy, has long since been left behind. It is still not properly understood just why some parts of substrate and deposit allow epitaxy (given appropriate temperature and supersaturation), while others give a random or partially oriented polycrystalline deposit. Meanwhile, Jesser's new theory is an important step forward.

1. W.A. Jesser, *Mater. Sci. Engng.* 4, 279 (1969).

15

Origin for Epitaxy

Nature, **233**, 524 (1971)

A recent article[1] adds a further stone to an unusual but increasingly solid edifice built over the past few years by Distler and his colleagues of the Institute of Crystallography in Moscow. Distler claims to have established a hitherto unconsidered origin for epitaxy — that is, the growth of one crystal species on a crystalline substrate of a different species, in an orientation defined by that substrate. He proposes that electric fields emanating from the surface of the substrate control the orientation, size and density of the epitaxial growths. Impressive evidence for this was published some time ago in *Nature*: [2,3]: it was then reported that a sufficiently thin amorphous intermediate layer between substrate and deposit failed to inhibit epitaxy, and furthermore the power to induce epitaxy could be 'imprinted' on the amorphous layer (selenium in this instance) by irradiating it before removing it from the substrate. The imprinted amorphous layer, by itself, supported an epitaxial growth of anthraquinone. These results were met by understandable scepticism and at least one investgator failed to reproduce Distler's early findings on the NaCl/amorphous C/gold system, but subsequently Henning[4] did confirm the findings.

Distler points out that all ionic crystals have charged point defects and assemblies of defects which can under certain circumstances — photoexcitation being particularly favourable — induce substantial permanent dipole moments and consequential fields at the surface. In effect, the crystals contain electrets, the electrical analogue to magnets. Just how the electric fields control the nucleation and growth of epitaxial layers — especially when they are electrically conducting — Distler does not attempt to propose, but it must be admitted that the circumstantial evidence is increasingly compelling.

In the latest article, Kobzareva and Distler make use of a recently established technique to deduce the sign of the local electric field from the local size, alignment and packing density of epitaxial anthraquinone crystals on triglycine sulphate (TGS), NaCl and mica (this time without an amorphous barrier). TGS is ferroelectric and so it would be expected readily to 'radiate' strong local fields. It was found that on each substrate, two kinds of deposit morphology were found, associated (it is supposed) with positive and negative fields. This hypothesis is reinforced by the direct observation, some time ago, of electrical heterogeneity on a mica surface, using an electrometer, whereas similar heterogeneity was established by an electron microscope technique applied to NaCl[5]. The implication of this latest study, taken together with the earlier ones, is that not only the nucleation but also, to some extent, the growth habit of the epitaxial deposit is controlled by electric fields emanating from the substrate. Distler insists, as he has done in his earlier work, that this makes epitaxial growth strikingly similar to some biological replication process. It now remains to discover how electric fields manage to do it!

1. S.A. Kobzareva and G.I. Distler, *J. Cryst. Growth* **10**, 269 (1971).
2. Distler and Zvyagin, *Nature* **212**, 807 (1966).
3. Distler and Obronov, *Nature* **224**, 261 (1969).
4. C.A.O. Henning, *Nature* **227**, 1129 (1970).
5. G.I. Distler, *Fiz. Tverd. Tela* **11**, 547 (1969).

16

New Light on Epitaxy

Nature, **239**, 254 (1972)

A recent paper dealing with epitaxy concerns a process which the authors, Metois, Gauch, Masson and Kern, quaintly christen 'post-nucleation'; what they mean by this is that epitaxy can be 'nucleated' after actual crystal nuclei have already been formed on the substrate[1]

Only one thing can be said with complete conviction about the phenomenon of epitaxy: it is not simple. Remarkable and sometimes contradictory observations abound; for instance, Russian work described in these columns last year[2] suggested, in disagreement with some earlier findings, that electric fields emanating from a substrate can influence the orientation of evaporated metal films. The central questions about epitaxy are these: what physical forces constrain a newly deposited crystal to align itself in a particular orientation relative to a monocrystalline substrate, and by what mechanism do these forces express themselves? Why just that orientation and no other, for a particular combination of deposit and substrate? Why does the orientation often change as experimental variables (including the nature of the crystal species, temperature and deposition rate) are changed?

If the past few years have seen considerable progress in answering at any rate some of these questions, much of the credit is due to the remarkable work done by Masson, Kern and their collaborators at the Université de Provence in Marseilles. Their most recent publication maintains the high standards of elegance, directness and clarity that one has come to expect.

The investigators cleaved KCl substrates in ultrahigh vacuum, immediately vapour-deposited aluminium or gold on the (100) cleavage face, heat-treated the film *in situ* and characterized the deposits by electron diffraction (to determine orientation) and electron microscopy (to determine the size distribution of crystallites). By comparing size distributions before and after heat-treatments, and by examining the migration of crystallites beyond the original sharply defined confines of the film, the investigators were able to monitor the exact behaviour of the crystallites, which were extremely small (typically some tens of Å in diameter).

For each deposited metal, at low temperature, a [111] fibre texture was formed: this happened because the crystallites originally deposited (presumably) in fully epitaxial orientation, and then migrated and underwent 'Brownian rotation', while all the time maintaining $(111)_{metal}$ parallel to $(100)_{KCl}$. The distribution of crystallite remained invariant. At higher temperatures, crystallites impinged on each other during their migrations and merged — but there was no classical 'Ostwald ripening'. By some as yet unexplained mechanism, the crystallites also rotated in the process of merging, so that for many grains $(100)_{metal}$ was then parallel to $(100)_{KCl}$ and also $\langle 011\rangle_{metal}$ was parallel to $\langle 011\rangle_{KCl}$ —that is, there was full epitaxy, not just a fibre texture. At intermediate temperatures, both forms of epitaxy coexisted, and selected-beam electron microscopy showed that the unmerged small crystallites retained the [111] texture while the larger crystallites acquired (100) epitaxy. All this indicates that both forms of epitaxy were instituted after the crystallites were already in existence — hence the term 'post-nucleation'. The authors comment pointedly that it is not reasonable to seek a single theory of epitaxial nucleation.

These elegant experiments still cast no light on why (111) epitaxy is preferred at low temperatures and (100) at high temperatures, and how and why the crystallites rotate to a new epitaxy as they merge by combination of smaller colliding crystallites. Nevertheless, this work takes understanding of epitaxy a good way forward.

1. Metois, Gauch, Masson and Kern, *Thin Solid Films* **11**, 205 (1972).

2. See Article 15.

17

Mobile Nuclei and Crystallite Coalescence

Nature Phys.Sci., **241**, 1 (1973) & *Nature* **246**, 66 (1973)

The boundary conditions that govern the deposition of thin metal films in epitaxial orientation on a monocrystalline substrate are still the subject of ingenious experimentation and protracted argument. Recent work in the laboratory of Masson and Kern at Marseille has focused attention on the role of processes that happen after the initial nuclei have already formed: Masson and Kern obtained incontrovertible evidence that some at least of these nuclei can migrate and rotate on the substrate surface and in the process the orientation relationship can change. To express this they coined the term 'post-nucleation" to indicate this source of epitaxy[1].

In a recently published conference report, Hirth[2] critically compares various theories concerned with the development of thin films, both those theories dealing exclusively with the nucleation stage and those embracing also the subsequent stages of growth and agglomeration, and accepts the important role of the later stages in understanding the overall growth kinetics and also in understanding the origins of epitaxy. He pays special attention to Ostwald ripening (the growth of some particles at the expense of smaller neighbours) of discrete nuclei of different sizes and discusses how the growth or contraction of a nucleus can be governed by the sizes of other nearby nuclei.

This same topic of mobility and coalescence of nuclei forms the subject of a stimulating paper in the same issue of the *Journal of Crystal Growth*[3]. Studying gold deposited on rock salt they show how the number density of nuclei passes through a maximum as a function of deposition time even though the mean size of nuclei continues to increase. At all stages, however (even after a long post-deposition anneal), a proportion of small nuclei obstinately survives. From an analysis of their observations, the investigators conclude that "the mobility of nuclei is both size- and orientation-dependent, and that the mobility of a particle formed by coalescence may be governed by the relative orientations of the combining nuclei as well as their orientations with respect to the substrate."

Donohoe and Robins discuss their findings in terms of a set of calculations by Reiss concerning the mobility of small nuclei on a crystal surface[4]. According to Reiss, small nuclei can migrate and rotate with activation energies of the same order as the diffusion barrier for single atoms on the surface, and further, any nucleus of arbitrary orientation can only migrate in certain limited directions; thus when two nuclei of different orientations coalesce. none of their "easy directions" may overlap and mobility ceases. This is one immobilization mode which may serve to interpret the survival, immobile and unabsorbed, of some relatively small nuclei even after long anneals. In general, Reiss's theory and the observations of Donohoe and Robins are consistent with the French group's conclusion that the acquisition by some nuclei of certain epitaxial orientations after nucleation coincides with a loss of mobility. It may well be that we are now due for a sudden burst of interest in Ostwald ripening as it happens in partly grown thin films, just as in the past decade the same phenomenon has been profitably studied with respect to small particles dispersed in three dimensions inside alloy grains.

Some theoretical implications of the Marseilles findings concerning two-dimensional migration and coalescence of crystallites have now been worked out by Robertson[5] in a stimulating paper. He starts from the observation (by the Marseilles team and others) that whereas crystallites do in fact coalesce at elevated temperatures (for example, above 150° C in KCl), they do not at lower temperatures, although other evidence indicates that they do migrate at these lower temperatures. Robertson analyses the predicted coalescence kinetics in terms of migration rates and collision probabilities (which themselves depend on crystallite sizes). It turns out that there should be extensive coalescence under conditions when in fact none is experimentally observed.

Robertson speculates on possible reasons why collisions should not result in coalescence. Three ideas are put forward: the crystallites may in various ways pick up electrostatic charges or the substrate surface may be contaminated; either form of disturbance is known to inhibit coalescence of colloids suspended in fluids. The third source of disturbance, examined in more detail, is the buildup of misfit stresses between substrate and crystallites, and Robertson indicates that the possibly sharp dependence of misfit stress on crystallite size could lead to either enhancement or hindrance of coalescence. He concludes by claiming that the very process of crystallite migration is not yet established beyond all doubt, and that this is a research topic deserving of much more attention, since it may force a complete revision of the classical concept of heterogeneous nucleation and growth.

1. See Article 16.
2. Hirth, *J. Cryst. Growth* **17**, 63 (1972).
3. Donohoe and Robins, *J. Cryst. Growth* **17**, 70(1972).
4. Reiss, *J. Appl. Phys.* **39**, 5046 (1968).
5. Robertson, *J. Appl. Physics* **44**, 3924 (1973).

18

Topotaxy in Metal-oxide Reduction

Nature, **316**, 297 (1985)

A thin metal film deposited by evaporation or sputtering on, say, a cleavage surface of sodium chloride is apt to be oriented in a reproducible way with respect to the substrate — this is the commonly found physical process of epitaxy, where the substrate is inert and unaffected. The chemical analogue is a topotactic reaction[1]: here a parent phase undergoes a reaction and the product phase has one or more defined orientations relative to the parent. Such reactions have been analysed by metallurgists (although they never use the term topotaxy) ever since the discovery of X-ray diffraction: the key investigations on so-called Widmanstätten reactions in metallic solid solutions were carried out in the early 1930s. Metallurgists usually analyse such reactions in terms of the fit of parent and product phases at their mutual interface.

The reverse reaction, reduction of an oxide to metal, has occasionally been examined from a kinetic point of view, especially in connection with the industrially important reduction of wuestite (FeO) to iron[2], but the first study of topochemical aspects has only just been performed. The results, reported by A. Revcolevschi and G. Dhalenne on page 335 of this issue of *Nature*[3], open up a new field of research.

In chemistry proper, topotaxy is usually demonstrated in dehydration reactions — for example, the conversion of $Mg(OH)_2$ into MgO — or the conversion of one oxide of iron into another; other instances include breakdown of complex silicates into simpler ones, with release of water[4]. In all these reactions, the observed orientations imply that the network of oxygen ions remains more or less intact throughout the course of the chemical reaction.

The most studied category of topotaxic reaction has been the oxidation of metals. The nature of both the topotaxy of a thin oxide layer and the oxidation kinetics depends not only on the oxidation conditions but also on the orientation of the metal surface (see refs 5 and 6 for reviews).

Revcolevschi and his collaborators at the University of Paris-Sud have for some time studied the orientation relationships between the phases of lamellar eutectistructures consisting of two metallic oxides (complementing the substantial body of information on such relationships as regards metallic eutectics[7]). for their new experiments, Revcolevschi and Dhalenne[3] have directionally solidified an eutectic of NiO and calcia-stabilized cubic ZrO_2, producing what is essentially an interpenetrating pair of epitaxially related single crystals which they then treat with a CO/CO_2 mixture that reduces the NiO to nickel without affecting the ZrO_2. The nickel proved to be in parallel orientation to the NiO which preceded it; the ZrO_2 phase acted as crystallographic reference. Just as earlier topotaxic experiments indi cated that the oxygen skeleton was maintained and unchanged, so the new experiment[3] indicates that the nickel-ion skeleton remains fixed.

This pioneering experiment opens the way to various topotaxic studies. One might try to mimic Gwathmey's celebrated experiment on the oxidation of a spherical copper monocrystal (showing the sharp anisotropy of oxidation rates and topotaxy) by reducing an oxide monocrystal phere, or to study the orientation dependence of reduction kinetics. It is known that the reduction rate of wuestite is sensitive to the population of point defects, and it would be interesting to see whether orientation relationships are affected by defects.

Revcolevschi and Dhalenne's observations have a potential practical interest: byreducing NiO without altering ZrO_2, it is possible to make a lamellar product where adjacent lamellae are successively electronically and ionically conducting. Such a structure can perhaps be applied in photoelectrolysis and in novel solid-state batteries.

1. J.M. Thomas, *Phil. Trans. R. Soc. (Lomd.) A* **227**, 251 (1974).
2. H. Schmalzried, *Solid State Reactions,* p. 195 (New York, Academic, 1974).
3. A. Revcolevschi and G. Dhalenne, *Nature* **316**, 335 (1985).
4. G.W. Brindley, *Prog. Ceram. Sci.* **3**, 1 (1963).
5. A.T. Gwathmey and K.R. Lawless, in *The Surface Chemistry of Metals and Semiconductors* (ed. H.C. Gatos) p. 483 (New York, Wiley, 1960).
6. J. Bardollé, in *Interfaces et Surfaces en Métallurgie* (eds. G. Martin *et al.*) p. 467 (Aedermannsdorf, Trans. Tech. 1975).
7. G.A. Chadwick, *Progr. Mater. Sci.* **12**, 99 (1963).

Crystals, Melting and Freezing

19

Phase Energies

Nature, **229**, 302 (1971).

As men living at the interface of the earth and air, as scientists living at the interface of the prosaic and the mysterious, as travellers scowling at customs men, we cannot escape interfaces. Certainly the detailed description of interfaces has been a sustained concern of scientists of many kinds; materials scientists, for instance, have long studied the morphology, composition and energies of interfaces between crystals or between phases. An exact knowledge of interfacial energies, in particular, is a precondition of understanding the behaviour of interfaces in systems undergoing transitions, and a particularly important category of transition is the homogeneous nucleation and growth of a stable phase at the expense of a metastable one.

According to Volmer, Becker and Döring's classical theory of homogeneous nucleation, an embryo of the stable phase becomes viable and grows steadily only when, by a chance fluctuation, it comes to exceed a critical spherical size. This size is such that a further atom, in deserting the metastable for the stable phase, releases in volume free energy more than the interfacial energy needed for the nucleus to increase in size. The theory has long been accepted as a central tenet of the theory of phase transformation (and of freezing of liquids in particular), but it has never been rigorously confirmed by experiment. The nearest approach, perhaps, was an attempt by Sundquist and Oriani in 1962 to relate critical undercooling in a two-liquid miscibility gap to the known interfacial energy. Subsequent criticism, however, cast doubt on the reliability of their conclusions. A. E. Nielsen and S. Sarig[1] have now published results, obtained by means of an ingenious technique, which really seem to constitute a quantitative confirmation of the Volmer-Becker-Döring theory. They studied nucleation of droplets of second phase in a ternary miscibility gap in the methanol-water-tribromomethane system. Two liquids were mixed rapidly and completely at a rapid flow rate, and the time τ to onset of turbidity was measured to a precision of a millisecond. This delay time constitutes an induction period for droplet nucleation plus diffusion controlled growth to a size large enough to scatter light to a detectable degree. Because the supersaturation of the mixed liquid is known and the volume fraction of new phase is small enough to have negligible effect on the composition of the parent liquid, all terms are known in the Volmer equations except the interfacial energy σ: thus σ can be calculated from the measured τ values for different compositions and supersaturations of the parent phase. Nielsen and Sarig compared these σ values with values measured directly by means of the classical drop weight method and obtained good agreement for the two sets of σ values, subject to a correction made for the variation of σ with surface curvature for very small droplets. This agreement was obtained for delay times less than 1 ms, corresponding to large supersaturations. Longer times corresponded to anomalous behaviour which Nielsen and Sarig attribute to heterogeneous nucleation.

Interfacial energies in another context have been considered by Bishop, Harth and Bruggeman[2]. These workers have examined the morphology of grain boundaries in zinc bicrystals of controlled orientation, and have found that for certain narrow ranges of mis-orientation (close to but not identical with twin orientations) the boundary became stably faceted — that is, it did not strive to achieve a state of minimum total area. The nature of the 'energy cusps' in the Wulff plot, which expresses grain boundary energy as a function of grain boundary orientation for a fixed mutual orientation of the contiguous lattices, was analysed in detail from the observed boundary morphologies: the determination of these cusps is an important achievement at a time when interest in grain boundary structure has suddenly revived after a long period of quiescence.

1. A.E. Nielsen and S. Sarig, *J. Cryst. Growth* **8**, 1 (1971).
2. G.H. Bishop, W.H. Harth and G.A. Bruggeman, *Acta metall.* **19**, 37 (1971).

20

Heterogeneous Nucleation from a Polymer Melt

Nature, **233**, 592 (1971)

The ingenious researches of David Turnbull in the nineteen fifties stimulated a great deal of work on the mechanisms and temperature dependence of nucleation in supercooled metallic and molecular melts. The classical Becker-Volmer theory of homogeneous nucleation is now firmly established for these categories of materials, but at the same time it has become clear that nuclei form homogeneously only under very exceptional circumstances: normal melts contain specks of impurity that act as nucleation catalysts. *A fortiori* , crystals forming from polymer melts would be expected to be heterogeneously nucleated always, in view of the extreme improbability that a sufficient length of polymer chain should align spontaneously in the correct configuration to constrain further growth in the proper pattern. But so far no heterogeneous nucleation catalyst for a polymer has been properly characterized and, generally speaking, this aspect of polymer science is confused.

In next Monday's *Nature Physical Science* , S. Y. Hobbs casts some light on this problem[1]. Hobbs crystallized spherulites of isotactic polypropylene from the melt, using short lengths of carbon fibre as nucleation catalysts. These were standard 'Morganite' fibres made from a polyacrylonitrile precursor, and two fibre variants were used — high strength (type 1) and high modulus (type 2). Type 1 fibre is known, from electron microscopy and small-angle X-ray scattering, to consist of very small turbostratic graphite crystallites of random orientation, whereas type 2 contains larger, highly oriented crystallites, with the basal graphite planes parallel to the fibre axis. Hobbs found that type 2 fibres were highly efficient nucleation catalysts, whereas type 1 fibres were ineffective.

The polymer chains in the crystalline structure of polypropylene are arranged in straight spirals. Hobbs showed that such a spiral can be accurately keyed to a graphite basal plane through some of the hydrogen atoms which (according to recent molecular orbital calculations) are preferentially chemisorbed at the midpoints of C-C bonds in the graphite structure. For effective nucleation the spiral has to be 'set', perhaps along one complete repeat period, and this requires the graphite basal plane which serves as substrate to exceed a certain minimum dimension, which Hobbs estimates at 50 Å. In type 2 carbon fibres, the constituent crystallites are more than 100 Å across measured in the basal plane, whereas in type 1 they are only -≈ 25 Å across. The inefficacy of type 2 fibres is thus readily understood. Type 2 carbon fibres have a crystallographic 'fibre texture', which implies that the normals to the basal plane are randomly distributed in a plane normal to the fibre axis. It would be interesting to continue Hobbs's work with single crystals of graphite or alternatively with highly oriented pyrolytic graphite. On the basis of Hobbs's interpretation of his observations, polypropylene would be expected to crystallize on such a substrate in two-dimensional spherulites, all polymer spirals being constrained to lie parallel to the exposed graphite basal plane. This form of nucleation would generate a unique form of preferred orientation with, presumably, strongly anisotropic properties.

1. S.Y. Hobbs, *Nature Physical Science* **234**, 12 (1971).

21

Crystal Defects and Melting

Nature, **273**, 491 (1978)

Of all the common properties of crystals, melting has a claim to be considered the most mysterious. Although there is no dearth of hypotheses, we do not know with certainty what structural features different crystal species have in common at the instant of melting.

In recent years, two venerable but mutually incompatible theories have been taken out and dusted, and their current champions each claim to have established strong evidence in their respective favour. The first is the dislocation theory of melting, first propounded by J. K. Mackenzie and N. F. Mott in 1952. The basic idea is that the energy of a dislocation array is a function both of dislocation concentration and of temperature (if dislocations are close together they partly relieve each other's stress fields) and if the temperature is high enough then a sudden catastrophic increase in dislocation density is predicted, to the point at which all crystallinity is lost. The early work is summarised by A. R. Ubbelohde[1].

R. J. M. Cotterill, W. D. Kristensen and E. J. Jensen have carried out extensive computer simulations, in two dimensions, which have shown that as kinetic energy is progressively injected, increasing populations of dislocation loops are created, grow and collapse again. When the instantaneous population grows large enough, the expanding loops "literally cut the crystal structure to pieces"; point defects are created in large numbers where moving dislocations intersect, and the combination of liquid-like disorder at dislocation cores and large point defect concentrations destroys crystalline order at a sharply defined effective temperature. The most recent account of these authors' work is in the *Philosophical Magazine* [2] and Cotterill has related this model to a number of other models of melting in a paper contributed to the Third Nordic High Temperature Symposium (1972).

The chief defect of this simulation approach is that it does not enable one to rationalise (let alone predict) the melting temperature at all accurately, although the model has been shown to be geometrically consistent with Lindemann's (empirical) Rule — that at the melting point, the root mean square dynamical displacement of atoms reaches a critical value: Cotterill *et al.* have shown that at a certain displacement, dislocation loops should spontaneously form. In this sense, an approximation to a melting criterion has been achieved, and to the extent that Lindemann's Rule is valid, the empirical facts are consistent with the dislocation theory in its newest variant. The model does to a first approximation interpret the volume change on melting in terms of the density changes at a dislocation core[3] and latent heat can also be interpreted[4].

An entirely distinct interpretation of melting, empirically beautifully self-consistent but devoid of mechanistic interpretation, is the vacancy model for metals, first advanced nearly 50 years ago by Frenkel and espoused recently by T. Górecki. In a series of papers, in English and Polish, he has adduced very extensive evidence that many different metals melt when the equilibrium vacancy concentration reaches a critical value of 0.37% and that there is then a sudden creation of further vacancies (or 'holes', in Frenkel's terms) to about 10%, at the cost of latent heat. The self-consistency of the model is tested in a number of ways. In an early paper[5], Górecki has shown how changes in various physical properties on melting can be rationalised: for instance, the increase in electrical resistivity on melting is proportional to the increase in resistivity per mole of vacancies for a range of metals, and the latent heat relates as expected to the formation energy of vacancies, E_v.

In another paper[6], Górecki has shown from a consideration of known radii of first coordination spheres in solid and molten metals, that the hypothesis of 9-10% vacancies in the melt at the melting temperature is tenable. A particularly interesting treatment[7] shows that the initial slope of the plot of melting temperature against hydrostatic pressure can be quantititively interpreted in terms of the pressure dependence of the formation energy E_v. This is a particularly convincing argument for the self-consistency of the model.

In a very recent paper, Górecki [8] answers some criticisms and also points to recent Russian work on several metals which shows that gamma irradiation (which of course creates point defects) lowers the melting temperature by an amount which is proportional to the dose (and therefore by implication to the additional vacancy concentration). In further papers, to appear in *Zeitschrift für Metallkunde,* Górecki quantitatively interprets changes in thermal conductivity and surface energy on melting in terms of the vacancy model.

As an empirical exercise, Górecki's model is watertight: he has built up a very wide range of distinct phenomena which it can interpret. What it lacks is a mechanism: he nowhere explains why and how the vacancy concentration on melting increases from 0.37% to about 10%. The dislocation model offered a mechanism to account for the volume change, but that model is not (on the face of it) consistent with the observed fact that all metals melt at about the same vacancy concentration.

A model might conceivably emerge from the consideration that vacancies are believed to become more diffuse as a metal is heated — that is, an isolated vacancy at room temperature becomes a region of reduced atomic density near the melting point. This was suggested by the classical simulation experiment of Fukushima & Ookawa[9]: they vibrated a liquid bath on which floated a 'crystalline' raft of identical soap bubbles representing atoms, and the vacancies became steadily more diffuse as the effective temperature increased. Groups of initially isolated vacancies merged into indistinguishable low density regions. The bubble raft is a very imperfect analogue of a three-dimensional crystal (though the raft did 'melt' when vibrated hard enough), but the motion of diffuse vacancy groups suggests regions of locally anomalously high vibrational amplitude, which might in turn trigger the destruction of crystalline order at a particular vacancy density. This might prove to be one way of answering the objection that a vacancy concentration of 0.37% is too low in itself to affect the stability of a crystal lattice. But that would still not provide any clue to the physical factors that determine the increase in vacancy (or hole) con centration in the melt itself.

A halfway position between metallic crystal and melt is occupied by a metallic glass. Such glasses contain holes just as do liquids, and a recent theoretical study [10] has established a way to estimate sizes and concentrations of such holes in different metallic glasses. Such glasses are found to have a total 'free volume' at the glass transition, T_g, which is a function of hole formation energy, E_h; but as the ratio of free volume at T_g to that at T_m is constant, hole size also varies with E_h and the vacancy concentration at T_m is constant, it would seem to follow that the hole concentration (as distinct from free volume) at T_g is approximately constant. This would seem to establish a parallelism between the glass transition and melting in terms of the vacancy model, but what has just been said also points to a possible weakness in Górecki's model, in that he tacitly assumes that holes in a molten metal are of the same size and have the same physical characteristics as vacancies in the corresponding solid; this approximation may well be invalid.

1. A.R. Ubbelohde, *Melting and Crystal Structure* (Oxford University Press, 1965).

2. R.J.M. Cotterill, W.D. Kristensen and E.J. Jensen, *Phil. Mag.* **30**, 245 (1974).
3. R.J.M. Cotterill *et al.*, *Phil. Mag.* **30**, 229 (1974).
4. R.J.M. Cotterill *et al.*, *Phil. Mag.* **27**, 623 (1973).
5. T. Górecki, *Z. Metallkde.* **65**, 426 (1974).
6. T. Górecki, *Z. Metallkde.* **67**, 269 (1976).
7. T. Górecki, *Z. Metallkde.* **68**, 231 (1977).
8. T. Górecki, *Scripta metall.* **11**, 1051 (1977).
9. Fukushima and Ookawa, *J. Phys. Soc. Japan* **10**, 970 (1955).
10. P. Ramachandrarao, B. Cantor and R.W. Cahn, *J. Mater. Sci.* **12**, 2488 (1977).

22

Melting and the Surface

Nature, **323**, 668 (1986)

Although freezing is a complicated pcess, there is general agreement about its essential features; but melting is still a mystery, and opinions about its origin abound. Recent experiments, however, seem to narrow the options and it is hard to avoid the conclusion that normal melting is initiated by a continuous vibrational lattice instability at the solid surface or at a solid–solid interface.

Ubbelohde[1], in a survey of models of melting, stresses the Lindemann criterion, formulated in 1910, according to which melting is a vibrational instability released when the root-mean-square amplitude of vibration reaches a critical fraction of the interatomic distance. In a comprehensive recent survey, Boyer[2] concludes that a lattice shear instability is the precipitating feature for melting, whereas Cotterill[3] emphasises catastrophic defect (dislocation) generation as the crucial factor. (I have discussed[4] the role of vacancies in melting in a previous *Nature News and Views* article.) Couchman and Jesser[5] made an early attempt at a synthesis of these apparently contradictory models: these authors pointed out that "the dislocation theory of melting, in essence the idea that [T_m is] the temperature at which the free energy of formation of dislocations is identically zero, implies, as does the Lindemann hypothesis, the vanishing of the static shear modulus". What is most important about their paper, however, is the recognition that melting is a surface-initiated process.

One clear indication of the role of the surface is that until recently it was believed to be impossible to superheat a crystal above its melting point except by arranging to heat a solid from the interior[6], so that the interior reached the melting temperature before the surface did; even then, the attainable superheating was always much smaller than the supercooling that can be imposed metastably on a melt.

This idea has now been disproved by an experiment[7] in which small (≈0.15 mm diameter) silver monocrystal spheres were coated with gold, as T_m (Au) is greater than T_m (Ag), and superheated for periods of the order of 1 min by up to 24 K above T_m (Ag); the limited time results from diffusion of gold from the coating into the silver. This experiment is different from that of Williamson *et al.* [8] who used pico-second electron diffraction in association with pulsed laser heating to show that thin, solid aluminium films can be superheated by more than 1 000 K but only for periods of $\leq$ 1 ns; here there is only transient but not metastable superheating.

Another body of evidence that assigns a crucial role in melting to the solid surface relates to the change of melting temperature with particle size. Buffat and Borel[9], among others, showed that small particles of Au, Pb and other metals suffer a melting-point depression of up to as much as 30 per cent for particles of 20-30 Å diameter. Hoshino and Shimamura[10] have interpreted this in the light of the well- known fact that the mean square amplitude of vibration at a solid surface is usually 1.5 – 2 times larger than in the bulk (for example, ref. 11); their theory is based on the idea that the Lindemann criterion for a vibrational catastrophe applies to the surface vibrational amplitudes rather than to the bulk values. Similar ideas have been suggested simultaneously by Pietronero and Tosatti[12].

Another kind of small particle is discussed by Rossouw and Donnelly[13], who injected argon ions into aluminium at room temperature. Under these circumstances, the Ar forms minute bubbles (mean diameter 2.6 nm) in the form of a crystalline face-centred cubic solid in epitaxy with the Al lattice. The melting temperature of argon at atmospheric pressure is 84 K, but allowing for the pressure in the bubbles (estimated from the measured anomalously small lattice parameter of the argon) it should be about 250 K. In fact, the diffraction pattern of solid argon is detectable up to 730 K, that is, a metastable superheating of 480 K!

Precision measurements of diffracted intensities allow mean square vibration amplitudes to be estimated for both Al and Ar, and the striking fact emerges that the Ar amplitudes are smaller than for free solid Ar. Specifically, the Debye temperature, a measure of cohesion deduced from the measured intensities, is 139 K for the Ar bubbles compared with 70 K for Ar at atmospheric pressure and 110 K computed from 'free' Ar pressurized in the bubbles. (A higher Debye temperature implies a higher cohesion, smaller vibration amplitude and, generally, a higher melting temperature.) Because the bubbles are so small that more than half the atoms can be regarded as being effectively surface atoms, it follows that the amplitude of surface (more precisely, interface) vibrations is depressed when Ar is in contact with Al (which has a higher Debye temperature of 398 K). Superheating is possible because the surface (interface) Ar atoms are constrained to smaller vibrations than the interior atoms. No doubt a similar explanation can be advanced for the substantial superheating observed for 2-nm hydrogen bubbles in an α–silicon matrix[14].

In a study similar to the argon-in-aluminium investigation, Evans and Mazey[15] injected krypton into titanium at 300 K and again found that the Kr was present in the bubbles as crystals epitaxial with the titanium, this time with a hexagonal close-packed structure. The Kr melts progressively during heating, the smaller bubbles (with smaller Kr lattice parameters and accordingly higher pressures) melting at higher temperatures. The melting of individual bubbles can be detected by high-temperature transmission electron microscopy, and survival of the smallest bubbles to higher temperatures is thus confirmed. The maximum superheating here is much smaller than in the case of Ar: the highest T_m observed is 380 K compared with a pressure-adjusted 'free' T_m of 335 K (the T_m for free Kr at atmospheric pressure is 117 K).

In commenting on earlier work by Evans and Mazey[16] on Kr bubbles in Cu and Ni, Rossouw and Donnelly[13] remark that the Kr lattice parameter increases (that is, the pressure decreases) after cycling through the melting transition and back; they presumed that this is caused by an inflow of vacancies from the host metal to the molten Kr bubbles. Such an inflow would modify the interface structure. They add that no corresponding change in the lattice parameter of their Ar bubbles in an Al host is observed after a similar melting cycle. This suggests that these bubbles are immune to vacancy inflow and the character of the interface is thus unaltered on cycling through the melting transition. Rossouw and Donnelly's experiment seems to be an unambiguous demonstration of the role of the surface (interface) in controlling the melting process. (In the study[7] of the melting of Au-coated Ag microspheres, the Ag was not under hydrostatic pressure, and therefore the modest superheating in that experiment was not connected with pressure.) The Ar-in-Al and Au-on-Ag experiments suggest that the observed superheating depends on epitaxy between the contained solid and the coating/host metal, and it is possible that that an incoherent interface might not permit substantial superheating.

Further evidence concerning the surface role in melting is provided by ion-channelling studies[17] using proton sensing beams on an atomically clean (110) surface of a heated lead crystal. The solid-liquid transition begins 40 K below the equilibrium bulk T_m, and the thickness of the molten surface layer increases dramatically close to the bulk T_m, being as great as 20 layers within a fraction of 1 K of that temperature. This implies second-order (continuous) melting at the surface, and thus raises in general terms the question of the order of transformation of the melting process at a surface; Frenken and van der Veen[17] quote recent theoretical work[18] on atomic ordering at a surface (for example, of Cu_3Au) which proves that an order– disorder transformation that is first-order in the bulk can be second-order (continuous) at the surface. That prediction has been confirmed by several experimental studies of Cu_3Au. Indeed, Frenken and van der Veen claim, but do not prove, that the Lipowsky/Speth approach[18], when applied to melting, predicts a liquid-film thickness diverging proportionally to $[T_0 / (T_m - T)]$, where T_0 is constant, which is consistent with their own experimental results for lead.

Thus, taking all the new evidence together, the almost unavoidable conclusion is that melting does indeed begin as a continuous vibrational instability at the solid surface of a solid--solid boundary.

1. A.R. Ubbelohde, *The Molten State of Matter: Melting and Crystal Structure* (Chichester, Wiley, 1978).
2. L.L. Boyer, *Phase Transitions* **5**, 1 (1985).
3. R.M.J. Cotterill, *J. Cryst. Growth* **48**, 582 (1980).
4. R.W. Cahn, see Article 21.
5. P.R . Couchman and W.A. Jesser, *Phil. Mag.* **35**, 787 (1977).
6. R.L. Cormia, J.D. Mackenzie and D. Turnbull, *J. Phys. Chem.* **65**, 2239 (1963).
7. J. Däges. H. Gleiter and J.H. Perepezko, *Materials Research Society, Symp. Proc.* **57** 67 (1986).
8. S. Williamson, G. Mourou and J.C.M. Li, *Phys. Rev. Lett.* **52**, 2364 (1984).
9. Ph. Buffat and J.-P. Borel, *Phys. Rev. A* **13**, 2287 (1976).
10. K. Hoshino and K. Shimamura, *Phil. Mag. A* **40**, 137 (1979).
11. R.M. Goodman and R.M. Somorgai, *J. Chem. Phys.* **52**, 6325 (1970).
12. L. Pietronero and L Tosatti, *Solid State Comm.* **32** 255 (1979).
13. C.J. Rossouw and S.E. Donnelly, *Phys. Rev. Lett.* **55**, 2960 (1985).
14. J.B. Boyce and M. Stutzmann, *Phys. Rev. Lett.* **54**, 562 (1985).
15. J.H. Evans and D.J. Mazey, *J. Nucl. Mater.* **138**, 176 (1986).
16. J.H. Evans and D.J. Mazey, *J. Phys. F* **15**, L1 (1985).
17. J.W.M. Frenken and J.F. van der Veen, *Phys. Rev. Lett.* **54**, 134 (1985).
18. R. Lipowsky and W. Speth, *Phys. Rev. B* **28**, 3983 (1983).

23

A Final Limit to Superheating

Nature, **334**, 17 (1988)

Metastability is the crucial feature of all phase transformations. At the exact thermodynamic equilibrium temperature between two phases, the rate of transformation is nil, and only by superheating or supercooling into the range of metastability can the transformation be driven at a finite rate. Even so, kinetic hindrance arising from slow atomic diffusion can brake or entirely prevent a transformation, even when the thermodynamic driving force is substantial. The amount of superheating or supercooling with respect to the thermodynamic transformation temperature which is feasible in practice is therefore defined by kinetic considerations. On page 50 of this issue of *Nature* , Fecht and Johnson[1] ask whether kinetics are the only relevant consideration, or whether there is such a thing as an absolute metastability limit, a temperature above (or below) which a metastable phase is forced to transform. They find that entropy gives an upper limit to the superheating possible for crystalline materials.

Fecht and Johnson[1] start with a long- established curiosity in the theory of glasses, the Kauzmann paradox[2]. Kauzmann pointed out that if the entropy of a liquid, plotted as a function of temperature, is extrapolated into the temperature range in which the liquid congeals as a glass (where configurational entropy is in fact frozen) then generally the extrapolation shows a temperature, well above 0 K, at which the entropy of the extrapolated liquid equals that of the crystal (see figure).

Kauzmann went on to advance reasons why, if a glass could maintain its configurational equilibrium down to the critical or 'Kauzmann temperature' at which its entropy equals that of the crystal, it would be forced to crystallize instantly. This is a paradox because at such low temperatures atomic diffusion is extremely sluggish so that crystallization should be impossible. Of course, as Kauzmann pointed out, the paradox is actually a metaphysical one, because it would be kinetically impossible for a glass to maintain configurational equilibrium down to such a low temperature. But the Kauzmann paradox continues to fascinate glass theorists and they continue to discuss its implications[3,4].

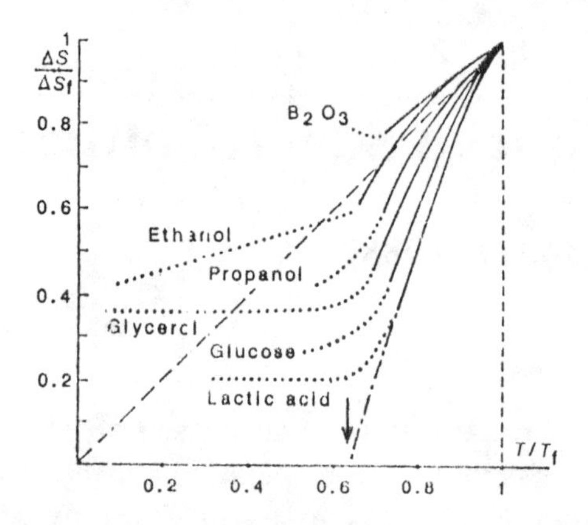

Variation of entropy (normalized with respect to the entropy of fusion) with temperature (normalized with respect to the freezing point). Solid curves, liquids in equilibrium; dotted curves, glassy state. Arrow, isentropic temperature for lactic acid.

It has long been recognized that, whereas a liquid can be metastably super-cooled, sometimes through hundreds of degrees, so long as heterogeneous nucleation can be discouraged (for example, by dividing up the melt into tiny discrete droplets), normally a crystal cannot be metastably superheated by more than a fraction of a degree. It can be superheated briefly, of course, while latent heat flows into the solid, but such a kinetic transient has nothing in common with metastable superheating. More recently, it has been recognized that melting, like freezing, is usually nucleated heterogeneously, at surfaces or interfaces (see my News and Views article[5]), and when special tricks are employed to prevent this then substantial superheating is feasible, for instance with noble-gas crystallites embedded epitactically in a metal crystal.

Fecht and Johnson discuss in their paper[1] whether there is an intrinsic limit to such superheating, just as the Kauzmann paradox sets an intrinsic — albeit metaphysical --- limit to the supercooling of a glass. They follow Kauzmann in seeking to locate an isentropic temperature, T_i^S , at which the entropies of the superheated crystal and the liquid phase are equal.

Fecht and Johnson calculate the entropyof superheated crystalline aluminium. It turns out that the role of lattice vacancies is crucial. Their concentration increases with rising temperature; if their contribution to entropy is neglected, then T_i^S coincides with the melting point, T_m, and superheating would be excluded. Taking the configurational entropy of vacancies into account yields an isentropic temperature (which is also the crystalline instability temperature) of 1 292 K = $1.38T_m$. In an illustration in Fecht and Johnson's paper, both upper and lower isentropic temperatures appear, the latter being the one associated with the Kauzmann paradox.

If the vibrational entropy associated with vacancies is also allowed for, T_i^S comes out a little lower, at 1 234 K. An isenthalpic temperature (1 217 K) can also be calculated; above this temperature, melting is no longer heat-flow limited. Indeed, it is hard to see what can prevent a crystal from spontaneously melting in the narrow temperature range between the isenthalpic and the isentropic temperatures.

The vacancy concentration at the instability temperature is found to be as high as 6-7 per cent. This calls to mind earlier experimental and theoretical results (mostly due to Górecki; see myearlier News and Views article[6]) suggesting that melting is correlated with the achievement of a critical vacancy concentration (taken to be less than 1 per cent by Górecki). But the effective vacancy (hole) concentration in an equilibrium melt at the melting point is estimated at around 10 per cent, and one way of regarding the isentropic catastrophe at $1.38T_m$ is that the solid becomes unstable against spontaneous collapse because of the high vacancy concentration. Fecht and Johnson express this by pointing out that at this temperature, the specific volumes of the liquid and of the superheated crystal become identical.

The authors conclude their stimulating *jeu d'esprit* by extending their instability concept from a pure metal to a solid solution, and present a hypothetical entropy diagram which shows T_i^S as a function of solute concentration.

The notion of what is in effect an upper absolute instability limit for the existence of a crystalline structure is a novel one: perhaps the closest parallel is the concept of an upper absolute instability limit for a first-order order-disorder transition[7].

1. H.J. Fecht and W.L. Johnson, *Nature* **334**, 50 (1988).
2. W. Kauzmann, *Chem. Rev.* **43**, 219 (1948).
3. M. Goldstein, *Ann. New York Acad. Sci.* **279**, 68 (1948).
4. L.V. Woodstock, *Ann. New York Acad. Sci.* **371**, 274 (1981).
5. See Article 22.
6. See Article 21.
7. A.G. Khachaturyan, *Progr. Mater. Sci.* **22**, 1 (1979).

24

New Ideas for the Melting Pot

Nature, **342**, 619 (1989)

Melting is distinct from other phase transitions in that it is not clear which of several mechanisms is responsible for initiating the process as temperature is raised. On page 658 of this issue of *Nature* , Tallon[1] joins a debate on what effect each of the possible precursors to melting may have on the extent to which crystals can be superheated.

Following my discussion three years ago[2] of the first observations of crystalline superheating above the equilibrium melting point, and the intimate linkage of this phenomenon with the nature of the crystal surface, Maddox[3] described new, strong experimental evidence that melting is indeed initiated at a crystal surface. Last year, Fecht and Johnson[4] (with myself acting as chorus[5]) proposed that, under circumstances that permitted super heating at all, there must be an upper limit, thermodynamically determined, to how far superheating can go. The central idea was that an 'entropy catastrophe', beyond which the crystalline entropy would exceed that of the liquid phase, would determine this limit. A few months later, Lele *et al.* [6] proposed an alternative way of estimating the temperature of the upper entropy catastrophe (there is also a lower catastrophe, the Kauzmann temperature, that marks the limit of superooling of a liquid/glass). It is this same question — the best way to estimate the upper limiting temperature tor superheating — that Tallon also addresses.

The crucial role of the surface in initiating melting has been underlined in several recent papers. The important work by Pluis *et al.* to which Maddox[3] drew attention has now been more fully described[7]: as the melting point is approached, a layer of disordering gradually develops; its thickness grows very rapidly within a few degrees of the melting point. Pontikis and Sindzingre[8] reviewed the tenuous evidence linking the surface roughening of crystals to surface melting and concluded that the two phenomena are probably entirely distinct. A bevy of theorists headed by Lipowsky (who has done earlier distinguished work in this field) has pursued the analysis of surface melting, and its anisotropy (which is a feature of the results from Pluis *et al.*) in terms of critical theory[9]. They introduce a family of novel order parameters to govern the melting process which depend on features of both crystal and liquid — thus beginning to address the old objection that most theories of melting look at the free energy of the crystalline phase but ignore that of the liquid altogether. Lipowsky's approach to melting readily interprets the role of the surface in nucleating the process.

Stoltzke *et al.* [10], who specialize in computer molecular dynamics simulations, added to the growing evidence that the surface layer acquires a very high concentration of vacancies (several per cent) just before it melts — a feature that Fecht and Johnson[4] took into account in their theory of an upper entropy catastrophe.

The main innovation in Tallon's new treatment is that he reintroduces into the analysis of upper and lower limiting temperatures for the crystalline and liquid/gas states, respectively, one of the oldest ideas in discussions of melting — Born's postulate[11] of a 'rigidity catastrophe', a temperature at which a shear modulus in the crystal falls to zero so that the crystal cannot sustain itself. This notion has been discounted of late, because real crystals do not lose their rigidity completely at the melting point. Using recent Monte Carlo calculations as a starting-point, Tallon shows that Born's idea is correct after all if the temperature of zero rigidity is taken to be the freezing point rather than the melt ing point (he argues that these points do not coincide in temperature-pressure-volume phase space). He demonstrates that an upper (absolute) instability limit for the crystalline state of aluminium, defined by the vanishing of the crystal rigidity, is at a temperature lower than the instability limits defined by an entropy catastrophe[4,6] and by an isochoric catastrophe (which he also treats) at which the specific volumes of crystal and liquid become equal. The rigidity catastrophe, Tallon claims, is what imposes the real upper limit to which a crystal can in principle be superheated.

There is just one feature of Tillon's treatment that is faintly disturbing. Instead of using the straightforward curve of free energy against temperature for the liquid, he subtracts the communal entropy of the liquid; this means that each atom is now constrained to remain within an unchanging set of neighbours or, in other words, the liquid is treated as a glass of infinite viscosity. It is not really clear why this is thought apposite for a temperature range in which the liquid in fact is bound to be highly mobile.

We now have three papers[1,4,6] that deal with the behaviour of crystals superheated through hundreds of degrees until they come up against a `no further' sign. Perhaps the time is now ripe for experimentalists, following the pioneers[12,13] to explore further the conditions under which various crystalline species can be metastably superheated, so that an experimental test of the various theories of the upper absolute instability limit will perhaps become feasible.

1. J.L. Tallon, *Nature* **342**, 658 (1989).
2. See Article 22.
3. J. Maddox, *Nature* **330**, 599 (1987).
4. J. Fecht and W.L. Johnson, *Nature* **334**, 50 (1988).
5. See Article 23.
6. S. Lele, P. Ramachandrarao and K.S. Dubey, *Nature* **336**, 567 (1988).
7. B. Pluis, J.W.M. Frenken and J.F. van der Veen, *Phys. Scripta* **T198**, 382 (1987).
8. V. Pontikis and P. Sindzingre, *Phys. Scripta* **T198**, 375 (1987).
9. R. Lipowsky, U. Breuer, K.C. Prince and H.P. Bonzel, *Phys. Rev. Lett.* **62**, 913 (1989).
10. P. Stoltzke, J.K. Nørskov and U. Landmann, *Surf. Sci.* **220**, L693 (1989).
11. M. Born, *J. Chem. Phys.* **7**, 591 (1939).
12. J. Däges, H. Gleiter and J.H. Perepezko, *Materials Research Society Symp. Proc.* **57**, 67 (1987).
13. C.J. Rossouw and S.E. Donnelly, *Phys. Rev. Lett.* **55**, 2960 (1985).

25

Dropping Below Freezing Point

Nature, **349**, 736 (1991)

By letting droplets of pure molten rhenium fall through a 48-metre evacuated tube, workers at Grenoble have succeeded in cooling this high-melting-point metal almost a thousand kelvin below its melting point before it solidified. The point of this study by Vinet *et al.* [1] was to probe the limits to undercoooling and to investigate the nucleation that precedes crystallization.

One can hardly become a materials scientist without encountering one of the central classics of this discipline, Turnbull's study of the limits of undercooling of molten metals[2,3]. When a molten metal is cooled, the amount by which it can overshoot the equilibrium melting temperature is normally determined by the presence of nucleation catalysts (an elegant name for minute specks of refractory dirt). Such specks ensure heterogeneous nucleation, caused by the existence of a low interfacial energy between a frozen nucleus and the speck. This always happens at a smaller undercooling than does true homogeneous nucleation, which is a purely statistical process in which a population of unstable solid embryos perpetually forms and remelts until the temperature falls far enough for a few to become stable and grow.

Turnbull's genial notion was to bypass heterogeneous nucleation by dividing the melt into tiny discrete droplets, typically tens of micrometres in diameter, surface–coating them to prevent coalescence and either supporting them on a substrate[2] or suspending them in an inert, immiscible fluid (the emulsion technique)[3]. The dirt specks are then restricted to a proportion of the droplets, and the remainder can freeze homogeneously.

The early (substrate) experiments were done with a range of metals melting at temperatures ranging up to 1 828 K (palladium), but the most precise experiments were done by the emulsion method with very low-melting-point metals such as mercury and bismuth. The undercooling was studied by dilatometry, which gave anomalous signals when a number of droplets froze and thus contracted. This kind of research has continued unabated since 1950 but rarely with the difficult materials now being studied by Vinet *et al.* who use the two metals with the highest melting points of all, tungsten (T_m = 3 690 K) and rhenium (T_m = 3 450 K).

Turnbull and Cech's early substrate experiments indicated that the maximum achievable normalized undercooling observed by this technique was restricted to 0.18-0.27. Turnbull[4] analysed this in terms of the classical theory of homogeneous nucleation, in which the predicted nucleation rate is proportional to $\exp(-k\,\sigma^3 T_c^{\,2}/T\,[\Delta T\,]^2)$, where σ is the interfacial energy between liquid and solid, T_c is the equilibrium melting temperature, ΔT is the undercooling and k is Boltzmann's constant.

From the exponent, it was a reasonable assumption that the important independent variable was the normalized undercooling ($\Delta T\,/Tc$). Turnbull went further[4], showing that the interfacial energy between solid and liquid should be proportional to the heat of fusion, ΔH_f , of each metal (at least for close-packed crystal structures) and concluding that the limiting normalized undercooling at which homogeneous freezing becomes unstoppable should indeed be about 0.18. It is easy to show that the theoretical nucleation rate for a given metal varies very rapidly indeed over a narrow temperature range: typically, a 10 per cent increase in undercooling will increase the homogeneous nucleation rate a million-fold — many experiments have confirmed this — and thus the critical undercooling can be quite narrowly defined.

In the early experiments, a modest standard cooling rate of around 0.3 K s^{-1} was used. Subsequent work showed that varying the cooling rate alters the achievable undercooling somewhat: a careful study by Chu *et al.*[5] on tin showed that the limiting undercooling increased from 95 to 103 K as the cooling rate is increased from 0.05 to 1.0 K s^{-1}.

Many experiments have been done with the substrate and droplet-dispersion approaches, and improved techniques (such as encapsulation in molten salts) permit increasingly high melting points to be examined. The emulsion technique in general yields greater undercoolings than does the substrate technique. The beginnings of microgravity research stimulated other methods, such as electromagnetic or acoustic levitation of droplets and the use of evacuated drop-tubes, and it became clear that such contactless methods often allowed undercoolings greatly to exceed those predicted by Turnbull's theory. Presumably, the absence of coatings and the cleanliness of the surface are responsible for this.

On the basis of this later work, some of it his own, Perepezko[6] proposed a modified limit to the possible undercooling, as much as

$0.65T_C$. Clearly, Turnbull's theory predicting a linear relation between interfacial energy and the latent heat of fusion needs critical reexamination: it works well, but only for σ values deduced from substrate experiments. The graph reproduced here (from one by Zarzycki *et al.* and including some recent results[7] on slowly cooled Si and Ge) gathers most data on undercooling of droplets and shows systematic variations according to the technique used.

Experimentally determined limiting undercoolings of molten metal drops, using the substrate technique (•), the emulsion technique (o) and contactless techniques (×), modified from Zarzycki *et al.* [8]. The solid line has the slope 0.18, predicted by Turnbull's theory.

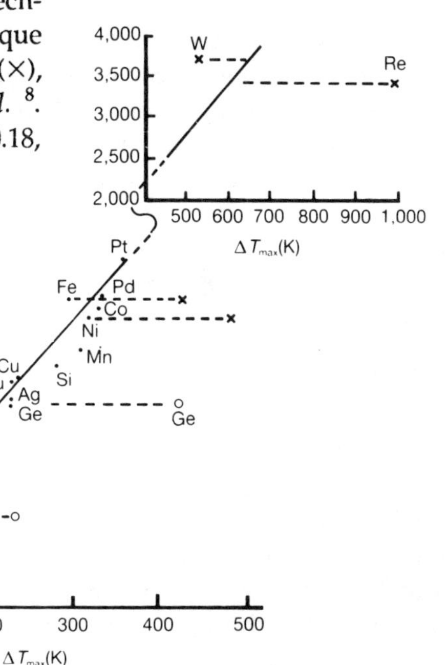

A possible source of the experimental deviation from simple theory is a wider than expected variation in liquid/solid interfacial energy, σ. This energy is hard to measure directly. Glicksman and his colleagues[9,10] developed a method centred on measurement of the groove formed where a grain boundary impinges on the solid/liquid interface. For bismuth, the measured interfacial energy[11] differed modestly, by 15 per cent, from that predicted by Turnbull's ΔH_f correlation. For lead, however[9], there was excellent agreement between the directly measured value of σ and that deduced from measurements of limiting undercooling by a variant of the substrate method; this constitutes one of the most convincing checks on the classical theory of homogeneous nucleation[12]. In their impressive new work on tungsten and rhenium, Vinet *et al.* used the drop tube technique. Metal drops were allowed to fall freely under gravity down an evacuated tube and their freezing *en route* detected by the associated recalescence (a transient reheating due to the release of latent heat). Recalescence was detected by one of a number of radiation detectors arranged along the length of the tube, and the temperature at the moment recalescence began was calculated from the position of the drop, the theory of radiative cooling and the initial temperature of the molten drop. The Grenoble drop tube is 48 m long and kept at an ultrahigh vacuum of 4×10^{-10} mbar. The cooling rate is high, of the order of 1 000 K s^{-1}, but as we have seen, this should make only a modest difference, a few tens of degree, to the achievable undercooling if the nucleation is homogeneous.

Fifty drops of tungsten, made by electron-beam melting the end of a wire, with 150 K initial superheat, were allowed to fall down the tube: ten showed recalesence with a mean undercooling of 530 K, equivalent to $\Delta T / T_c$ of 0.14. The range of ΔT among the ten drops was small, 28 K, as expected from classical theory. The other forty drops presumably froze heterogeneously, not surprising when it is realized that the drops were over 4 mm in diameter. Ten rhenium drops, of about the same size, again at 150 K initial superheat, were tested, and all showed high undercooling: this time the mean ΔT was 975 K, with a very small spread (only 0.5 per cent). This corresponds to a remarkable value of $\Delta T / T_c$ of 0.28.

The authors point out that the extremely good vacuum ensured the cleanliness of the falling drops and diminished the likelihood of heterogeneous nucleation: this, together with the strikingly narrow range of ΔT observed for those drops which recalesced, makes it highly probable that true homogeneous nucleation had been observed. The undercooling found for rhenium is the highest absolute value observed for any metal hitherto, and the values of σ deduced for tungsten and rhenium are also the highest known for any metal. From the graph, however, it can be seen that the undercooling for tungsten (W), large as it is, is smaller than predicted by extrapolation from lower- melting-point metals: perhaps nucleation was heterogeneous after all.

Vinet *et al.* are planning to examine a range of other high-melting-point metals, such as platinum, tantalum, osmium and their alloys, by the same technique; when they have done this, it will be clearer how consistent the behaviour of these refractory metals is with that of low- melting-point metals.

1. B. Vinet, I. Cortella, J.J. Favier and P. Sesré, *Appl. Phys. Lett.* **58**, 97 (1991).
2. D. Turnbull and R.E. Cech, *J. Appl. Phys.* **21**, 804 (1950).
3. D. Turnbull, *J. Chem. Phys.* **20**, 411 (1952).
4. D. Turnbull, *J. Appl. Phys.* **21**, 1022 (1950).
5. M.G. Chu, Y/ Shiohara and M.C. Flemings, in *Chemistry and Physics of Rapidly Solidified Alloys* (eds. B.J. Berkowitz and R.O. Scattergood) p. 3 (New York, Met. Soc. AIME, 1983).
6. J.H. Perepezko and J.S. Paik, in *Rapidly Solidified Amorphous and Crystalline Solids* (eds. B.H. Kear, B.G. Giessen and M. Cohen) p. 49 (New York, Elsevier, 1982).
7. P.V. Evans, G. Devaud, T.F. Kelly and Y.-W. Kim, *Acta metall.* **38**, 719 (1990).
8. J. Zarzycki, G.H. Frischat and D.M. Herlach, in *Fluid Sciences and Material Science in Space* (ed. H.U. Walter) p. 599 (Berlin, Springer, 1988).
9. G.E. Nash and M.E. Glicksman, *Phil. Mag.* **24**, 577 (1971).
10. R.J. Schaefer, M.E. Glicksman and J.D. Ayers. *Phil. Mag.* **32**, 725 (1975).
11. M.E. Glicksman and C.I. Vold, *Acta Metall.* **17**, 1 (1969).
12. K.F. Kelton, in *Solid State Physics* (eds. H. Ehrenreich, F. Seitz and D. Turnbull) Vol. 45, p. 75 (Academic, New York, 1991).

Mechanical Properties of Materials

26

Clean Break?

Nature, **218**, 67 (1968)

Fundamental Phenomena in the Materials Sciences. Volume 4: Fracture of Metals, Polymers and Glasses. (Proceedings of the Fourth Symposium, held Jan. 31- Feb. 1, 1966, at Boston, Mass.). Edited by L. J. Bonis, J. J. Duga and J. J. Gilman. 310 pages. (New York: Plenum Press, 1967). $ 16.

How hard must I pull this piece of solid to make it break ? At first sight, nothing could be more straightforward than this question; its meaning appears obvious, and the answer a matter of the simplest of experiments. In fact, the question is subtle in the extreme, the answer hedged about with qualifications, and the interpretations varied and disputed. The question is subtle because all sorts of variables need to be specified: simple tension or complex stress system; monocrystalline, polycrystalline or amorphous material; rate of loading; size distribution of cracks, grains, polymer chains; plastic deformation mechanisms available; temperature. The answer is statistical, and even statistical reproducibility is the prize for extreme care in standardizing variables. The interpretations are divided into two fundamentally distinct categories: the microscopic approach, concerned with mechanisms on the atomic scale, and the fracture mechanics approach, in which the solid is treated as a continuum with specified properties — as a black box, so to speak.

The interdisciplinary conference, the record of which is here under review, was convened partly to bridge this duality of approach to the problem of understanding fracture, and partly in the hope that the very different behaviour of crystalline materials and of polymers might result in mutual illumination. The exercise was well worth while, and the book makes most instructive reading, not least for the long, discursive discussions which reveal metallurgists, continuum mechanics specialists and polymer physicists groping to grasp each other's concepts and preoccupations.

Where crystalline materials are concerned, and polycrystalline alloys in particular, the microstructural approach has yielded a rich harvest of understanding, and correlation with the engineering approach has advanced far. Hahn and Rosenfield's 'systems approach' to fracture of steels is a striking illustration of this correlation. (An instructive comparison can be made with the largely unsuccessful attempt at another conference, twelve years ago, to marry the microscopic and continuum approaches to plastic deformation.) The continuum approach to polymer fracture has also become quite subtle, but microscopic analyses do not seem to have passed far beyond the descriptive stage; the two approaches have not yet a great deal to contribute to each other. Nevertheless, Rosen's chapter on the generation of microfissures in viscoelastic polymers (analogous to microcracks in metals) is particularly stimulating, and points the way to a wide open field of research. Rosen and McMahon (who writes on the microstructural aspects of tensile fracture of metals) agree that vast numbers of tentative microfissures or microcracks form first and progressively develop to form the large crack that eventually causes fracture. There is no clean break; polymers, indeed, when beset by microfissures, warn of their impending destruction by visibly 'blushing', a form of early warning to which politicians unfortunately seem immune.

The metallurgists at the conference, accustomed to the primacy of the microstructural approach, appear to have been somewhat nonplussed by the clear lead, in the case of polymers, of rigorous analysis on the basis of purely empirical relationships between strain, stress, time and temperature. Professor Ansell's remark, in discussion, that "*structure* in a metal is relatively simple, as compared with the polymers, while the *behaviour* in the metals is much more complex" certainly seems justified to me, trained as a metallurgist. Dr Temin, a polymer man, reacted to this view somewhat acidly: "There seems to be a tendency for those who are most learned in metals to presume the polymer situation to be extremely simple", and Dr Bonis pointed out that we all believe the other man's field to be much easier than our own. His hope that the polymer/crystal dialogue should be helpful has undoubtedly been justified, and the book deserves the attention of fracture specialists of all persuasions.

27

New Forms of Strengthening

Nature, **222**, 416 (1969)

Dispersion-hardening is the standard strategy for strengthening intrinsically weak metals. Small particles of a hard phase are distributed uniformly in the metallic matrix, either by precipitation *in situ* or by sintering together. 'Duralumin', the 'Nimonics', thoria-disperse nickel (TD-nickel) and sintered aluminium powder (SAP) are examples of the genre. It has always been assumed that the metallurgist should aim to achieve the greatest possible uniformity in dispersion so as to optimize the resistance to plastic flow. N. Hansen, of the Danish Atomic Energy Commission, has now shown this to be unnecessary[1]. He has atomized aluminium powder, covered the particles with a thin layer of aluminium oxide (no technical details of preparation are offered), consolidated and extruded the material. The product embodies a three-dimensional network of oxide particles, a few hundred Ångströms in size, forming cells of the order of a micron in diameter.

One unusual feature of the material is that, unlike SAP and other uniformly dispersion-hardened alloys, the extruded alloy readily recrystallizes. The recrystallized alloy contains large grains partitioned into much smaller cells by a close elongated network of alumina particles. Slip planes cross many cells; the flow stress of the alloy varies with cell size according to a Hall-Petch relationship, which indicates that the network walls are effective in arresting the passage of dislocations. The slope of the Hall-Petch plot is in accord with the hypothesis that dislocations eventually break through the barrier by the Orowan mechanism. The efficacy of strengthening is almost as great as for uniform dispersion.

Some interesting work now needs to be carried out to assess whether the material will keep its strength at high temperature in spite of its readiness to recrystallize. SAP and TD-nickel depend at least in part for their good mechanical properties on the stable dislocation pattern produced during processing (see, for example, ref. 2[2]) and the retention of these dislocations at moderately high temperatures because recrystallization is inhibited. Hansen's new material may be less dependent on the presence of an artificially induced dislocation population and stand or fall according to the stability of the cell structure itself. It is also possible that a coarse-grained dipersion-hardened material will have advantages in applications, such as for canning nuclear fuel elements, in which good creep resistance is important.

Another category of strong materials currently exciting interest is the directionally solidified eutectic composite. The argument here is that it is always difficult to make good reinforced composites by consolidating a matrix with prefabricated fibres or whiskers because some damage to the latter is bound to result from the fabrication process itself. There is thus a clear advantage in making the reinforcement *in situ* : not only is damage avoided, but the bond between matrix and reinforcement is necessarily better. Yue *et al.*[3] adopted this strategy in making specimens of Al–$CuAl_2$ alloy of eutectic composition, directionally frozen so that a single eutectic colony is formed, its platelets aligned with the direction of heat flow. The samples were tested in compression in various directions, and found to be particularly good when tested parallel to the platelets. On yielding, the alloy buckles and forms well defined kink-bands in which the plates adopt different orientations. At 45° to the platelet plane, the compressive strength is at a minimum. The paper incorporates an analysis of the mechanics of a platelet structure deformed in compression.

The mechanical properties can be enhanced by heat treatment so as to precipitation-harden the matrix, which is supersaturated with aluminium; the material thus benefits from the combination of two quite distinct strengthening mechanisms.

1. N. Hansen, *Acta metallurgica* **17**, 637 (1969).
2. Guyot and Ruedl, *J. Mater. Sci.* **2**, 221 (1967).
3. A.S. Yue, F.W. Crossman, A.E. Vidoz and M.I. Jacobson, *Trans. Met. Soc. Amer. Inst. Min. Metall. Engrs.* **242**, 2441 (1968).

28

Strong Fibrous Solids

Nature, **225**, 592 (1970)

The growing interest and importance of strong fibrous solids were underlined by the Royal Society's discussion on this topic held in London on January 29. Mr W. Watt (Royal Aircraft Establishment, Farnborough) summarized the structural changes believed to take place in polyacrylonitrile (PAN) fibres during progressive heating in oxidizing or neutral atmospheres, with or without tensile stress, and the associated changes in properties at each stage. He emphasized that the statistical variability in tensile strength depended substantially on sample length, which indicates the importance of defects in limiting the strength. It is uncertain whether the vital defects are the inclusions, or the elongated pores recently established by diffuse X-ray scattering and electron microscopy. Watt stressed the paucity of knowledge about structural changes in PAN, for example, concerning the degree of crystallinity induced by drawing, and called for intensive investigation. The opinion was often expressed at the meeting that the technology of fibre composites has too far outstripped scientific understanding.

Dr R. J. E. Glenny (National Gas Turbine Establishment, Farnborough) recounted systematic experiments on the use of fibre composites based on refractory matrices specifically for use in gas turbine blades. On the basis of the criteria of density, strength and chemical compatibility, attention was focused on a nickel-base matrix with wires of tungsten or tungsten/ rhenium alloy, Mo-RTZ alloy or alumina. Vacuum-casting of blade shapes in a cored shell-mould gave excellently uniform filament distribution and it proved possible to incorporate clear channels for coolant. Tungsten and alumina were the most promising reinforcements. Strength levels were marginally less than predicted by the rule of mixtures, and fatigue behaviour was little affected by the presence of filaments. It is too soon to be sure of the future scope of these techniques, but in view of the fact that niobium and chromium alloys are not considered to be equal to expectations, filament-reinforcement seems to be the most promising way to improve on the intrinsic properties of nickel superalloys.

Dr J. Nixdorf (Battelle-Institut, Frankfurt) described the pilot-plant production of various strong metallic microwires, including wire of a marageing steel. The technique involves melting a metal rod held in a viscous glass mantle (which acts as both container and lubricant) and rapidly drawing the assembly through a die.Afterwards the glass covering is removed. The strength is attributed not only to the small diameter of a few microns but also to the rapid cooling. A particularly small statistical variability of the wire strength was claimed, and the mean strength exceeded that of boron but was less than that of E-glass. Little was said about stiffness. Various composites have been made with both metallic and polymeric matrices, and for concentrated composites a UTS better than for boron or carbon reinforcement was claimed. The production cost of microwires (on a scale resembling the present scale of production of carbon fibres) was assessed at some tens of pounds sterling per kg. Oddly enough, the audience made no reaction to these intriguing disclosures.

Describing the development of 'Hyfil' (carbon fibre-reinforced) compressor blades for Rolls-Royce jet engines, Mr J. I. Goatham (Rolls-Royce, Derby) emphasized the extensive testing procedures needed, both for development and for routine testing, including such exotic methods as holography. He also mentioned the great importance of joints and attachments, where critical losses of weight advantage can easily be sustained by faulty design. He was the only speaker to emphasize the joining problem, though the chairman, Dr A. H. Cottrell (Cabinet Office, London), expressed the view that this central problem was well on the way to being vanquished.

Dr F. J. McGarry (Massachusetts Institute of Technology) reported work on epoxy resin toughened by elastomer dispersions — following the example of rubber-reinforced polystyrene as described at the Royal Society in 1963. The impressive increase in fracture toughness which was achieved seemed to depend in part on achieving an optimal degree of cross-linking of the elastomer, as well as a bimodal distribution of coarse and fine elastomer particles. McGarry postulated that coarse particles favour crazing of the matrix, which absorbs energy, while small particles promote shearing of the polymer matrix, which again absorbs energy. The ultimate aim is to use the toughened polymer as a matrix for fibre-reinforced building materials. The potential of the building industry as a consumer of composites was stressed by Dr A. J. Majumdar (Building Research Station, Watford), who showed how concrete or plaster panels could be toughened by incorporating (through a co-spraying technique) fibres of special zirconia-bearing glass which can stand up to the corrosive alkaline environment in concrete (there is no special problem with plaster). Majumdar explained that plaster panels thus reinforced will stand up to repeated impact from primary school children.

Dr A. Kelly (National Physical Laboratory, Teddington) continued the theoretical approach by analysing the dependence of work of fracture on the length and diameter of fibres. It transpired, to the surprise of many, that from this point of view thick fibres were preferable. In brittle/ductile composites, thick fibres favour plastic deformation over large volumes of a ductile matrix, while in brittle/brittle composites, thick fibres suffer pull-out over greater lengths, with the increased associated frictional energy loss. Thus two physically quite distinct processes, between them, lead to a law of general validity for fibre composites.

Professor F. C. Frank (University of Bristol) assessed the ideal stiffness of a carbon chain by various means (infrared spectroscopy, calculations from the characteristics of diamond and so on) and was led to a value of 2–2.5 Mbar. The best value obtained in practice is apparently about ten times less than this, though it might have been predicted that hard-drawing, of nylon, for example, should reduce this factor of disparity. Frank went on to describe some special structures which seemed to consist of a stiff polymeric spine with built-in polymeric beads, which might open the way to a polymer-fibre-reinforced polymeric matrix. He predicted that quite soon the polymer industry, in search of higher strength, would turn to more elaborate production methods, even though this would diminish one of the chief advantages of polymers, their cheap fabrication.

29

Measuring the Invisible

Nature, 229, 526 (1971)

Fifty years ago a distinguished Rolls-Royce enginer, A. A. Griffith, postulated an explanation of the fact that the strength of brittle solids such as glass is surprisingly low and moreover statistically very variable. He proposed that the surface is pitted by many submicroscopic cracks which can locally amplify an applied stress.

Circumstantial evidence has long since established Griffith's hypothesis beyond all cavil. Practical improvements in the strength of glass and other brittle materials have been based on Griffith's recognition that the surface determines strength; strength can be increased by either removing the cracks through etching or otherwise and at once protecting the surface by means of a hard coating, or by prestressing a thin surface layer in compression.

Throughout all such technological developments, Griffith's microcracks have remained shadowy entities, apprehended but not seen. A number of methods have been tried to make them visible: plenty of cracks were revealed, but there are reasons for believing that the treatment may create more new cracks than it enlarges old ones.

Next Monday's issue of *Nature Physical Science* includes an article by J. D. Poloniecki and T. R. Wilshaw[1] which establishes a new way of tackling the problem. The technique is the Hertzian fracture test. A hard sphere (here a tungsten carbide ball 0.77 mm in diameter) is pressed into the glass surface under continuous observation until a ring crack suddenly spreads and grows concentrically with the indentation. The radius of this crack, r_c, and the load P at which it forms are recorded, and the test is repeated many times. The ring radius and critical load are related to the surface density of microcracks of various depths, and therefore in principle it is possible, from distributions of r_c and P, to deduce histograms of surface densities of cracks as a function of crack-depth. This has been recognized for some years, but the problem has been to find an acceptable way of analysing the data statistically. Most previous investigators have been content to postulate a particular form for the histogram (a 'Weibull distribution'), which prejudges the central issue. Moreover, nobody has proposed simultaneous analysis of the distributions of both r_c and P.

Poloniecki and Wilshaw's investigation is the outcome of a collaboration between a fracture mechanics expert and a statistician, and it now seems that the analytical problem has been solved. The critical step is to assess the surface area 'at risk'' in each individual test, and to sum these for all the tests: this total area at risk is different according to the particular range of crack depth under assessment, and has to be separately estimated for each bar of the histogram. The technique has been applied to mechanically polished silica glass and a log/log histogram produced, complete with 90 per cent confidence limits.

1. J.D. Poloniecki and T.R. Wilshaw, *Nature Phys. Sci.* **229**, 226 (1971).

30

Toughened Glass

Nature, **238**, 245 (1972)

It is one of nature's little ironies that many materials of particularly high strength are of little use as constructional materials. Substances such as silica, hard glass, alumina and boron nitride have extremely strong chemical bonding and thus very high intrinsic strength as well as high softening temperatures, but are also so brittle that they cannot safely be used under large loads, except in pure compression; also the risk of fracture by mechanical or thermal shock is unacceptably great. There are promising exceptions — notably silicon nitride — but glass and ceramics have usually been disappointing as materials for the mechanical engineer. This is why recent attempts to toughen these materials are specially important.

Reinforcement of weak materials with strong and stiff fibres or whiskers is now well understood in the laboratory and engineers' wariness is gradually diminishing. In industrial practice, all such composites have to date been based on polymeric matrices, most work having been carried out on glass-fibre reinforced epoxy resin. Carbon fibre is also being used to reinforce resins, but carbon-fibre reinforcement of metals for high-temperture use is less promising because of chemical attack of the fibres.

The very recent development of a radically new way of making continuous monofilaments of sapphire[1,2] offers new prospects of making effective high-temperature composites based on metal matrices. Here the high strength of the filaments will be utilized and the metal matrix will serve merely as a binder to transfer stress to the fibres and (because of its deformability) to inhibit the spread of cracks.

A rival strategy is to use a strong brittle matrix with strong brittle fibres, and to depend on the interaction between these two brittle constituents to generate toughness. This is the idea underlying recent work at Harwell on the reinforcement by carbon fibres of glass, alumina, magnesia and glass-ceramic[3,4]. Sambell and his colleagues prepared their materials by hot pressing, with either discontinuous fibres or aligned continuous fibres, up to 50 per cent volume fraction. Substantial increases in bend-strength were recorded with the continuous fibres but there were variable results with discontinuous fibres. The fracture toughness, however, increases for all matrices and all fibre arrangements; in the most favourable combinations the increase in the work of fracture was more than a thousand-fold. For discontinuous fibres, the work of fracture increased with fibre content up to 30 per cent; beyond this, no further benefit was normally obtained. For continuous fibres, however, a sharp increase in work of fracture came at about 50 per cent volume fraction, provided the method of pressing used was designed to prevent excessive damage to the fibres. The results were best for glass matrices, least good for magnesia.

Sambell *et al.* discuss their very promising findings in terms, especially, of the mismatch of thermal expansion coefficients of matrix and fibres; this mismatch leads to internal stresses and for some matrices, especially magnesia, leads to matrix cracking during cooling from the hot-pressing temperature, and such materials have relatively poor mechanical properties. Quite generally, however, matrix cracking occurred well before the ultimate tensile strength of the composite was reached.

Sambell *et al.* assume that the very high fracture toughness of some of their materials stems entirely from the work required to pull broken fibres out of the matrix, as cracks spread through the matrix. They ignore the work of fracture of the matrix itself, even though this must be far from negligible if the matrix is subject to multiple cracking. Sambell's analysis is cast in doubt by some work by Cooper and Sillwood of the National Physical Laboratory[5].

Cooper and Sillwood studied a model system consisting of epoxy resin reinforced with regularly spaced continuous steel wires of various diameters and volume fractions. The matrix was caused to crack in a very regular pattern by pre-cooling the composite in liquid nitrogen. This technique was adopted for simplicity in preference to mechanical testing which, however, led to similar cracking. For most composite samples, the observed crack spacings so created varied with fibre radius and volume fraction V_f in a highly regular way: the crack spacing was proportional to $(1 - V_f)/V_f$, and this is readily rationalized on the hypothesis that the critical fibre load-transfer length determines the crack spacing.

The really interesting feature, however, was the behaviour of samples with a high volume fraction of very fine wires. Such composites could not be cracked ahead of final failure, by either thermal or mechanical treatment. In an effort to understand this, Cooper and

Sillwood worked out a Griffith-type energy equation, relating externally provided work to the total internal energy release, by fibre stretching, matrix cracking and frictional resistance to fibre withdrawal. Their equations show that for a high volume fraction of fine fibres, there is indeed insufficient external energy available to sustain cracking of the matrix. They point out that their equations allow a composite to be designed so that matrix cracking and fibre fracture initiate simultaneously instead of sequentially. The matrix will then not begin to crack until the normal breaking strain of the matrix is reached, but (because of the need to pull out the simultaneously breaking fibres) the toughness will be much higher than for the unreinforced matrix. It seems probable that the sudden sharp increase of work of fracture of a glass matrix observed by Sambell *et al.* as they increased the volume fraction of aligned carbon fibre to 50 per cent is to be understood in terms of Cooper and Sillwood's theory: the sudden increase may well correspond to the critical situation outlined earlier.

The glass/carbon and ceramic/carbon composites studied at Harwell maintained their strengths at high temperature if tested in an inert atmosphere, but lost strength in air because the fibres oxidized. The chemical reactivity of carbon fibres, which limits their use in hot metal matrices, is thus seen to limit their usefulness in glass and ceramics also; but it may be that methods can be found to protect the fibres against oxidation, or else perhaps useful scope will be found for the composites in non-oxidizing atmospheres.

1. T.J.A. Pollock, *J. Mater. Sci.* **7**, 631, 649, 787 (1972).
2. G.F. Hurley, *J. Mater. Sci.* **7**, 471 (1972).
3. R.A.J. Sambell, D.H. Bowen and D.C. Phillips, *J. Mater. Sci.* **7**, 663 (1972).
4. R.A.J. Sambell, A. Briggs, D.C. Phillips and D.H. Bowen, *J. Mater. Sci.* **7**, 676 (1972).
5. G.A. Cooper and J.M. Sillwood, *J. Mater. Sci.* **7**, 325 (1972).

31

Oscillating Yield in a Plastic

Nature, **251**, 15 (1974)

Some years ago, a group of Russian polymer scientists[1] discovered a novel form of behaviour in several amorphous polymers. The neck which forms during tensile straining of the polymer turned out to be made up of successive bands of translucent and opaque material, and the stress-strain curve showed corresponding strain oscillations. The opaque material had evidently crystallised.

This phenomenon of 'oscillating necking' has been theoretically analysed by Barenblatt in the Soviet Union[2] and to back up this analysis, Roseen of AB Atomenergi in Sweden has adopted an unusual experimental approach[3]. He examined a tensile specimen of amorphous (poly) ethyl terephthalate during tensile de formation by means of an infrared Thermovision camera, which records instantaneous temperature distributions. Strips of high temperature (up to 95°C) develop in association with the opaque crystalline strips. The calibration and the resolving power of this camera are set out in some detail. It seems, therefore, that during oscillating necking, a narrow layer of crystalline material forms internally (micrographically it was shown that the surface remains amorphous). Crystallisation seems to be the result of rapid local deformation, which adiabatically heats a narrow front of polymer. The crystalline material is much less ductile than the amorphous matrix; plastic flow and its associated heat generation slows down, the stress rises and the cycle repeats.

Thus, polymers may show behaviour superficially similar to the discontinuous yielding long familiar in certain alloys. This is known as the Portevin-Le Chatelier effect and is most familiar in mild steels. In essence, it is caused by the cyclic locking of dislocations by readily mobile impurity atoms (normally carbon or nitrogen). A group of dislocations breaks free, the impurity atoms catch up and lock them afresh, and here also the cycle repeats.

Roseen has confirmed the postulated difference in tensile

behaviour between amorphous and crystalline polymer by examining the effect of intentional heat treatment of the polymer on the form of the tensile curve. Here is another unfamiliar point of similarity between metals and polymers: it is second nature to a metallurgist to examine the effect of systematic heat treatments on the mechanical behaviour of an alloy, but this is not yet a normal experiment for a polymer scientist.

One lesson from this intriguing piece of work is that it can be illuminating for a polymer scientist to know something of physical metallurgy. But it would be hubris to leave it at that. Increasingly, metallurgists have something to learn from polymer scientists, as concerns both phenomena and experimental techniques. It is time for 'polymer crystallisation' to become a normal part of the syllabus of students of metallurgy and materials science.

1. Andrianova *et al.*, *J. Polymer Sci.* **9**, 1919 (1971).
2. Barenblatt, *Mekhanika Tverdogo Tela* (No. 5), 121 (1970).
3. Roseen, *J. Mater. Sci.* **9**, 929 (1974).

32

Revolutionary Silicon Carbide Fibre

Nature , **260**, 11 (1976)

G. K. Chesterton once wrote a rollicking poem about the Rolling English Road, and commented affectionately on the entertaining journey from somewhere–or–other to Birmingham, by way of Beachy Head. Anyone who knows England knows that it is not sensible to go to Beachy Head if you want to finish up in Birmingham, no matter where you start from. But it was an Englishman who had the crazy notion of making pure carbon (Birmingham) by first making an elaborate polymer (Beachy Head) and then destroying it by heat. Watt's technique of stretching, oxidising and pyrolysing polyacrylonitrile to create strong carbon fibre was highly unconventional, but it worked triumphantly. [Note: It turned out that this approach was in fact first used in Japan, and then further developed by Watt.]

Now an equally eccentric and equally brilliant Japanese chemist, Seishi Yajima, has found an equally unconventional way to make strong, continuous fibres of silicon carbide. A preliminary summary of his method has been published in *Chemistry Letters (Japan)* [1]. More detailed publication is on the way, and Yajima (who is head of the Oarai Branch of the Research Institute for Iron, Steel and Other Metals of Tohoku University) has also privately circulated an account of his underlying philosophy. He explains that conventional organic chemists have long striven to convert inorganic materials to organic (this goes right back to Wöhler's synthesis of urea), but they have an inbuilt resistance to turning an organic material, attained in complex form by considerable efforts, back into an inorganic material of simple structure by heat or chemical treatment: "The organic chemist intrinsically detests a high temperature. He cannot stand the destruction of a complicated structure which has been obtained by painstaking efforts." Yajima is an unconvent ional organic chemist who has overcome the phobia of fire. [Note: Yajima at this time encountered grave obstacles in his own country, of an undisclosed nature, to the adoption of his methods — which are now well established there — and died prematurely, apparently by his own hand.]

Fibre-reinforcement for service at high temperatures is fraught with difficulties. Carbon fibre is too reactive, alumina is too weak, SiC whiskers are too short and too expensive. Intrinsically, however, SiC has all the required properties, and Yajimi set out to make strong continuous fibre. He starts with dimethylchlorosilane, $(CH_3)_2SiCl_2$, reacts it with lithium and a catalyst to form dodecamethylcyclohexane, $[(CH_3)_2Si]_6$, which is polymerised in an autoclave. The molecular weight fractions are separated by repeated solvent extraction and a solution of molecular weight ≈ 1 500 is solvent-spun into a fragile, brittle fibre. This is then pyrolised in a vacuum at temperatures in the range 800–1,300°C to turn it into β SiC. Infrared spectroscopy and X-ray diffraction were used (the details are in process of publication) to follow the stages of pyrolysis. The best fibres (1 300°C heat-treatment) have strengths of 300-350 kg mm^{-2}, and elastic moduli of 30 Mg mm^{-2}. Diameters are typically 10-20 μm.

The extraordinary feature of the fibres is that although their strength matches that of SiC whiskers, their structure is totally different. Electron microscopy and analysis of X-ray diffraction line profiles both reveal a structure of minute cryatallites, 7 nm across, without preferred orientation. This fine structure is crucial: if pyrolysis is done at 1 500 °C, giving considerably larger crystallites, the strength falls sharply. Prolonged attempts were made to match this structure by sintering ultrafine SiC powders, but these were never fine enough to achieve high strength. Yajima can only speculate as to the reasons for the remarkable strength associated with the ultrafine morphology.

The SiC fibres are extremely resistant to heat and oxidation and will withstand, undamaged, prolonged exposure under load to a flame at 1 200°C. By contrast, SiC *whiskers* lose strength above 600°C. No information is given about the economics of the process, but it appears that the process can be made to work on an industrial scale.

Yajima claims that his work is the first instance of a metal atom being introduced into the skeleton of an organic polymer. This innovation has resulted in an inorganic product which appears at last to open the way to an effective fibre-reinforced composite, SiC-in-metal, which can give effective service at high temperature.

1. S. Yajima, *Chemistry Letters (Japan)* (No. 9) 931 (1975).

33

The Power of Paradox

Nature , **347**, 423 (1990)

Science progrsses by the resolution of paradoxes. Interpretations of the nature of light swung back and forth — particles or waves? — until the quantum synthesis embraced both absolutes. A door must be either open or shut, the logical French tell us. But a liquid crystal is neither solid nor liquid; a quasicrystal, neither crystalline nor amorphous; a magnetic 'fluid' can be either mobile or rigid. The best doors are both open *and* shut.

So it is with strength. Ceramics are much stronger than metals: but drop a plate on the floor and it shatters, drop an aluminium bottle and it dents and survives. What really matters is toughness, the ability to resist shock, to break gradually rather than catastrophically, to give adequate warning of impending disaster. Brute strength will not suffice. The introduction of fibre-reinforced composites — arrays of strong ceramic fibres in weak polymeric matrices — as a practical way of taming the brittleness endemic to ceramics expressed the principle of "strength made perfect in weakness", to cite an apposite biblical paradox. The fibres arrest cracks which try to spread through the polymer, and also carry most of the load; but polymers will not stand up to red heat.

Now, as reported elsewhere in this issue of *Nature* [1], a group from the ICI Advanced Materials Laboratory has found a new way of cresting toughness by marrying strength and weakness; their way produces a composite material which is relatively cheap to make and moreover contains only components resistant to high temperatures. The combination of intrinsic ceramic strength and heat resistance with both toughness and cheapness is something new.

Clegg and his collaborators have prepared blocks consisting of silicon carbide layers, separated by thin graphite interlayers. Graphite in itself is far from weak — in the form of carbon fibres, it is one of the strongest materials known — but here it forms a weak interface: a crack in the silicon carbide impinging upon the interface becomes diverted so that it carries on spreading along the interface, dissipating energy as it does so. Silicon carbide and graphite are both strong but the interface between them is weak in shear, not in tension or compression: strength made perfect in weakness. The material, in fact, behaves very much like nature's own favourite composite, wood.

The composite is made by mixing silicon carbide powder, doped to aid sintering, with a concentrated polymer solution in a special high-shear processor. The ICI team had previously developed both the solution and the mixer to make ultra-strong cement. The resulting silicon carbide paste, in the form of 2 mm sheets, is rolled out like dough into layers 0.2 mm thick, coated with graphite, stacked in tenfold multi-layers, slowly dried out and sintered. Graphite, we are told, has the crucial property of not reacting with stoichiometric silicon carbide, which can dissolve no excess of eitherof its constituents. For comparison, the authors also sintered unrolled, uncoated silicon carbide sheets.

Slow bend tests at ambient temperature show, for the composite, gradual fractur, with repeated load drops as incipient transverse cracks on the tension side are diverted to spread laterally along silicon carbide/ graphite interfaces. A specific work of fracture — one test of toughness — ranging from 5 600 to 6 700 J m^{-2} was recorded, compared with a value of only 60 J m^{-2} for the monolithic (non–composite) material (a figure further reduced by a technical correction to 30 J m^{-2}). The composite is thus more than a hundred times tougher than the monolith, at least in bending and at room temperature. The stress at which failure begins in the composite is reduced by 25 per cent, however: every benefit has its price.

The work of fracture of the composite is slightly higher than that of a hardwood, deal. So the ICI team has improved upon nature in two ways: their material is tougher than hardwood and it can certainly be safely heated to red heat. Whether the material maintains its strength and toughness at high temperature we are not told.

The new material is, in effect, a ceramic/ceramic composite. Others such have been made before, but only by expensive methods such as gas-phase reactions which take months to make what the ICI method produces in hours. Carbon/carbon composites[2] are an exception: they are an industrial product category in actual use, and are made by impregnating woven carbon fibre structures with liquid or gaseous polymeric or other organic precursors. However, they are mostly used for applications such as rocket nozzles or fighter wheel–brakes for which cost is scarcely considered. Very strong microlayered aluminium/transition metal composites have been made at the Royal Aerospace Establishment in Farnborough by vapour deposition: this kind of alloy is very strong and tough but seems for the time being to have faded from view, perhaps because of its exorbitant cost.

Clegg *et al.* [1] point out that their new technique is not restricted to making sheets. For instance, wires of silicon carbide/carbon composites can be extruded from the paste and then coated and pressed together to make rods. But the real novelty of the approach is undoubtedly the intrinsic cheapness of the processing method.

1. W.J. Clegg, K. Kendall, N. McN. Alford, T.W. Button and J.D. Birchall, *Nature* **347**, 445 (1990).
2. S.E. Hsu and C.I. Chen, in *Superalloys, Supercomposites and Superceramics* (eds. J.K. Tien and T. Caulfield) p. 721 (New York, Academic, 1989).

Order/Disorder

34

New Superlattices

Nature, **230**, 356 (1971)

The term *superlattice* was coined by an X-ray metallographer many years ago to describe a crystalline solid solution in which, after appropriate heat-treatment, solvent and solute atoms are deployed in an ordered manner. In strict semantic terms, the word only makes sense for alloys such as the Cu/Pd series in which 'long-period superlattices' occur: the crystal pattern has a structural discontinuity every l unit cells in one particular direction, where l is a small integer, typically 5. Such a superlattice can be described as one-dimensional. Only recently has the work of Sato and Toth at the Ford Scientific Laboratories provided an understanding, in terms of electronic band theory, for the existence of such superlattices.

Work in several unconnected fields of solid-state science has now established that long-period superlattices in alloys are not the only exemplars of the species. In recent months, new forms of one-dimensional and three-dimensional superlattices have been described, and a two-dimensional form was predicted and subsequently observed some years ago.

The new *one-dimensional* superlattice is a man-made semiconductor structure: a silicon monocrystal slice is grown from the vapour phase in an atmosphere which is rapidly and regularly alternated between phosphine-rich and arsine-rich compositions. The slice grows with plane, differentially doped layers at a regular spacing, typically about 150 Å[1]. Their work was stimulated by a remarkable theoretical paper by L. Esaki and R.Tsu published a few months earlier[2]. Esaki and Tsu worked out in some detail the band theory and electron transport properties of one-dimensional semiconductor superlattices. In effect, such materials contain mini Brillouin zones bounded by small energy gaps, all interpolated inside the principal Brillouin zone, and at the same time they embody periodic fluctuations of the main energy gap. The effective electron mass becomes strongly dependent on the direction of motion of the electron, and in the authors' words, this "leads to virtually a two-dimensional electron gas system".

An important theoretical consequence of this feature is that the drift velocity along the normal to the superlattice layers passes through a maximum as a function of applied electric field, and so at sufficiently high fields (not enough, however, to cause tunnelling and avalanching) the system has a negative differential conductance. In effect, this happens when the electronic mean free path sufficiently exceeds the superlattice period. Such systems, in which the host crystal and the scale, amplitude and nature of the dopant superlattice are independently disposable variables, should give rise to a novel class of devices and open new areas in semiconductor physics, a field which has lost its first fine rapture. The preliminary experimental work by Blakeslee and Aliotta has now shown that semiconductor superlattices can readily be made with the requisite period and amplitude and with a large number of layers, and so it seems that the new field is wide open.

The new *three-dimensional* superlattice is a regular array of voids in irradiated molybdenum, recently described by J. H. Evans[3]. Voids are small cavities in a crystalline lattice produced by neutron or particle bombardment. In molybdenum bombarded with 2 MeV nitrogen, electron microscopy showed that the voids were arranged on a regular body-entred cubic lattice aping that of the host metal, but with a repeat distance around fifty times larger (220±10 Å). This remarkable observation has now been explained by R. Bullough and K. Malen at a recent conference on voids at the University of Reading. They showed that superlattice formation by voids depends on elastic interactions between the voids, which in turn relies on the anisotropic elastic properties of the molybdenum crystal lattice; the theory is able to interpret the scale of the observed periodicity. Presumably the voids migrate by the well-attested process of self-diffusion along their surfaces until the optimum periodicity is established, but this has not been examined. The superlattice contains some vacancies and dislocations, though so far these have not been studied in detail. The nomenclature of a vacancy in a void lattice would seem to offer an intriguing exercise in double negatives.

The dimensional catalogue concludes with the *two-dimensional* flux-line lattices in type 2 superconductors, predicted in 1957 by Abrikosov and directly observed during the past four years. The lattice consists of normally conducting filaments carrying a magnetic field, and over a range of temperatures below the upper critical field these flux lines exist in equilibrium and arrange themselves in a two-dimensional lattice with either a triangular or a square unit cell. The important feature of the lattice (which typically has a repeat distance of about 500 Å) is that, in a hysteresis-free type 2 superconductor, the entire flux-line lattice is unstable the moment a finite current flows through a superconductor. The critical field in this temperature range is thus zero unless the lattice can be anchored in some way to the underlying crystal lattice.

A great deal of metallurgical research is currently being devoted to the study of anchoring to dislocations and to second-phase particles. Until recently, such studies had to depend entirely on indirect deductions based on measurements of critical currents and fields, but a technique invented by Träuble and Essmann in 1966 has made it possible to observe the flux-line lattice directly. Colloidal particles of iron are allowed to deposit on a superconductor surface, decorate the exits of flux-lines and become fixed in position, so that replicas can later be taken and examined by electron microscopy. A. Seeger has recently published a survey of superconductivity and physical metallurgy[4], in which much of the active recent research in this field is summarized and its strategy explained. It seems that flux-line lattices themselves contain all the standard defects — grain boundaries, vacancies, edge dislocations and even disclinations (a species of defect which, though specified by theorists such as Nabarro, has not been observed in crystal lattices). After illustrating all these defects in the flux-line lattice, Seeger goes on to discuss the physical nature of their interaction with defects in the crystal lattice itself and the consequence of such interaction for superconducting behaviour. This article is a particularly readable introduction to a subject which bristles with conceptual difficulties.

It now only remains to seek connections between these different kinds of macroscopic superlattices. Perhaps void lattices can be created in a periodically doped superconductor and unprecedentedly large critical fields thus obtained?

1. A.E. Blakeslee and C.F. Aliotta, *IBM J. Res. Dev.* **14**, 686 (1970).
2. L. Esaki and R. Tsu, *IBM J. Res. Dev.* **14**, 61 (1970).
3. J.H. Evans, *Nature* **229**, 403 (1971).
4. A. Seeger, *Metallurgical Trans.* **1**, 2987 (1970).

35

Ordering in Films

Nature, **239**, 50 (1972)

Most diffusion studies are performed on bimetallic couples so arranged that interdiffusion of the constituents produces no intermetallic compounds, for such phases greatly complicate analysis of the experiment. When one or more compounds do form in such a couple, the smooth gradation of composition which is characteristic of diffusion couples is interrupted by a sudden jump at the surface of this layer of the compound. A complex alloy system, such as the Cu-Zn system, results in a complicated sequence of intermediate layers, composition jumps alternating with composition gradients. The very considerable complexities involved in analysing such situations have frightened off many investigators. Only G. V. Kidson, who was concerned with the uranium-aluminium system, has really come to grips with the thermodynamics and kinetics of the situation[1]. It seems that all phases in the equilibrium diagram should be expected to turn up as layers in the annealed couple, but if the diffusion coefficients and homogeneity range for a particular phase are unfavourable, the corresponding layer may be so thin that it escapes detection under an optical microscope.

These considerations are brought to mind by an extremely interesting investigation just published by two staff members of IBM[2]. Their interest in the behaviour of bimetallic films arises from the fact that successive evaporation of two metals is routine in microelectronic fabrication. So they prepared thin ($\approx$ 0.8 μm) films by sequentially evaporating copper and gold. The films, before and after various heat-treatments at 200° C, were examined by means of a specially designed X-ray diffractometer using an oblique (non-symmetrical) focusing geometry which assured adequate sensitivity even with such a very thin film. The X-ray scans provided informationabout lattice parameters and intermediate phases present, and furthermore the sequence and thicknesses of layers were unambiguously established by depositing a second film in inverse sequence of the constituent metals, giving both films the same heat–treatment and comparing the normalized intensities from corresponding layers; these are different because of different attenuation by absorption in the two cases.

The conclusion emerged that the annealed films retained some *unalloyed* Cu and Au, separated by successive films of the ordered phases Cu_3Au and $CuAu_3$. There was no sign of the other ordered phase, CuAu. Both these features — the lack of a concentration gradient in the terminal metals, and the missing phase — are distinctly puzzling, but the authors do not comment on them. The 'missing' phase may well be there in the form of a layer so thin that it cannot be detected, or it may be that some nucleation barrier prevented such a layer from forming in the first place. The lack of concentration gradients, however, seems to be entirely inconsistent with the established laws of intermetallic diffusion. It is also most interesting that, for the purposes of a diffusion couple, an ordered structure such as Cu_3Au acts as a separate phase with a limited homogeneity range. This 'phase identity' of ordered structures is a matter on which much controversy has raged (see, for example, O'Brien and Kuczynski[3] and Irani and Cahn[4]), and this new observation is a further indication that ordered phases are to be regarded as classical Gibbsian phases. (It is, however, mystifying that $CuAu_3$ turns up as a separate layer, since 200°C coincides with the critical temperature for disordering of that phase.)

The rates of thickening of the Cu_3Au and $CuAu_3$ layers were determined from accurate intensity measurements of diffraction lines; growth was linear for both kinds of layer, and the rates were identical. This rate was much too fast to be due to volume diffusion and much too slow to be due to grain-boundary diffusion (in view of the extremely small grain size). The authors conclude that the growth rate could not have been controlled by the rate of diffusion of Cu and Au through the ordered layers, but must have been limited by the atomic mov-ments at the interface required to secure the correct order and stoichiometry.This conclusion does suggest that perhaps no CuAu layer was able to form at all.

1. G.V. Kidson, *J. Nucl. Mater.* **3**, 21 (1961).
2. K.N. Tu and B.S. Berry, *J. Appl. Phys.* **43**, 3283 (1972).
3. J.L. O'Brien and G.C. Kuczynski, *Acta metall.* **7**, 803 (1959).
4. R.S. Irani and R.W. Cahn, *Nature* **226**, 1045 (1970).

36

Homo or Hetero?

Nature, **271**, 407 (1978)

At first sight, an ordered metallic solid solution is as simple and perfect as a crystal can be. Solvent atoms sit on one kind of lattice site, solute atoms on another. The supposed simplicity and perfection are deceptive, for any ordered alloy differs from a plain intermetallic compound in that its ordered structure is unstable and dissipates on heating before the solid melts. The interest of such alloys does not lie in the perfection of the order, but in the mysterious intermediate states in which order begins to disappear; the halfway house between perfect crystallographic order and an utterly random distribution of atoms on the available lattice site. Are the wrongly sited atoms scattered at random among the 'right' ones — is there a randomness within the randomness? — or are there minute groups of 'wrong' atoms, right with respect to their immediate neighbours but wrong with respect to the main mass of the ordered crystal? The first is a homogeneous model of partial order, the second, a heterogeneous one. The homo/hetero dispute goes deep in this field of solid state physics.

When an ordered alloy is heated above its critical temperature, all long-range order disappears, but local or short-range order persists. Its persistence may be deduced from the observation of diffuse X-ray scattering in place of sharp superlattice lines, or from small thermal, electrical or dilatometric anomalies. The atomic format of short-range order is subject to the same uncertainty between homogeneous or heterogeneous models. One view is that a short-range ordered alloy is homogeneous in the sense that any solvent atom has statistically the same likelihood of having like or unlike nearest neighbours as any other solvent atom. Of the several heterogeneous models, the most intriguing postulates a dispersion of tiny fully ordered domains surrounded by a wholly random matrix with a thin boundary in which there is a gradient of the degree of order. In other words, ordered 'precipitates' exist in a disordered matrix. This model, associated with H.P. Aubauer, has occasioned a great deal of controversy since it was first advanced 6 years ago. The difficulty is this: in the normal way, any dispersion of precipitates in a matrix is unstable; the Matthew Principle operates and large precipitates grow at the expense of the small. The same happens with soap bubbles or with raindrops in a cloud. Aubauer[1] pointed out that a short-range ordered, heterogeneous alloy remains indefinitely in a staff of fine dispersion: the ordered microdomains do not grow however long they are heated. Aubauer attributes this to the presence of an elastic strain gradient around each domain (because the volume per atom is different in ordered and random regions), and the elastic energy is shown to inhibit the coarsening of domains. Aubauer's calculations have repeatedly been challenged on thermodynamic grounds (the story of the dispute has been neatly anatomised by Martin and Doherty[2].

The homogeneous/heterogeneous dichotomy with regard to short-range order seems to be incapable of final resolution by X-ray diffraction, electron diffraction, electron microscopy or electrical resistivity. Copper-rich, α–phase Cu/Al solid solutions, in particular, have been examined by all these techniques, over a period of 20 years, and though the partisans of heterogeneity have the edge, the arguments have been impossible to resolve decisively.

A deceptively simple investigation has now, to all appearances, tilted the balance firmly in favour of a heterogeneous model (although that still leaves open the question whether Aubauer's particular version is the most apposite). Trieb and Veith[3], two Viennese physicists working in consultation with Aubauer, have studied several Cu/Al alloys exclusively by measurements of electrical resistivity. In the past (for instance in a classic study by Wechsler and Kernohan[4]) this has been done by quenching from high temperatures, which leads to great complexities arising from quenched-in vacancy populations that lead to kinetic anomalies. Trieb and Veith avoided large temperature changes: an ultraprecise measurement technique allowed them to cool or heat their samples through only 5 deg C, so that in effect the only variable to affect the resistivity was a change in degree of short-range order. By approaching the same end-state from above and below, they were able to define the temperature range in which equilibrium was attainable in reasonable times in their various alloys. They then analysed mathematically the form of the kinetic curves and showed unambiguously that these could only be interpreted on the hypothesis that two different processes were in progress at different rates. Twin-process kinetics cannot be reconciled with a homogeneous end-structure. Trieb and Veith point out that in Aubauer's 'disperse order' model, an alloy of given composition has at each temperature, in equilibrium, a defined volume fraction of ordered domains and a defined mean size of domains. They propose that the equilibrium volume fraction is attained several times faster than is the equilibrium degree of dispersion, following any sudden small change of temprature. The electrical resistivity is sensitive to both variables and can thus be used to follow their changes. In the most concentrated alloy (18 at. % Cu) the disordered matrix virtually disappears and is replaced by antiphase boundaries.

Trieb and Veith have made a very convincing case for the correctness of a heterogeneous model and have shown how powerful a basically very simple technique, allied with extreme precision and careful quantitative analysis, can be.

It is intriguing that the homo/hetero dispute extends also to the structure of oxide glasses. An enormous amount of work has gone into analysis of such glasses by diffuse X-ray scattering, but it is fair to say that the resultant pair distribution functions cannot reliably discriminate between homogeneous (Zachariasen-type) network models and various models based on non-crystalline clusters, or clusters of microcrystals of two polymorphic variants in coexistence. This battleground of glass structures still awaits its resolution, which will perhaps come, in due course, from some technique as elegant and simple as that applied by Trieb and Veith to short-range order.

Their study has shown again the limitations of microscopy, however great the resolution and however sophisticated the technique, in resolving details of not quite perfect structures.

1. H.P. Aubauer, *Acta metall.* **20**, 165 (1972).
2. J.W. Martin and R.D. Doherty, *Stability of Microstructure in Metallic Systems*, pp. 205-209 (CambridgeUniversity Press, 1976).
3. L. Trieb and G. Veith, *Acta metall.* **26**, 185 (1978).
4. M.S. Wechsler and R.H. Kernohan, *Acta metall.* **7**, 599 (1959).

37

Superlattices, Superdislocations and Superalloys

Nature, **309**, 745 (1984)

Superalloys, the key materials for making turbine blades and discs in jet engines, have been steadily improved since the introduction of the wartime 'Nimonics'; maximum service temperature has been progressively raised and engine efficiencies have been improved in parallel. But the available means for maintaining this rate of improvement are now more or less exhausted, short of a radically new approach. Just such an approach is offered by some recent 'pure' research on one constituent phase present in all superalloys, the nickel-based 'gamma prime', or γ' phase.

The diverse family of superalloys is, in essence, based on a single structural principle. They consist of a dispersion of particles of a coherent gamma prime phase — Ni_3Al in its pure form but generally containing Ti and other elements — within a gamma phase, a disordered matrix of nickel containing various solutes. Gamma and gamma prime are coherent in the sense that the lattices match up so closely that their interface is of very low energy; this in turn implies that the gamma prime particles are highly stable against growth. The other vital characteristic of these alloys is that the gamma prime phase is atomically ordered; in other words, it consists of a superlattice, like the much-studied Cu_3Au alloy, but unlike in that alloy, the order persists undiminished right up to the melting temperature. All sorts of other microstructural features have been introduced to add to the excellent creep resistance of superalloys at high temperatures, but the above features are the crucial ones.

Gamma prime has a remarkable property: whereas all ordinary alloys weaken as they are heated, the strength of an experimental alloy of pure Ni_3Al gamma prime phase increases as it is heated. Plainly, this property is of central importance in determining the good creep resistance of superalloys as a class. Such anomalous mechanical behaviour is confined to Ni_3Al and a number of other A_3B phases, such as Ni_3Ti. Much ink has been spilt in attempts to interpret this anomaly. It is known that dislocations move through highly ordered alloys in pairs, collectively termed 'superdislocations'. In a very stably ordered alloy these pairs of dislocations, which are joined by strips of disordered material called antiphase domain boundaries, are so close together that they can twist and turn and escape on to abnormal lattice planes. It is the twisting and turning that leads to the anomalous mechanical behaviour; at least, this is the current orthodoxy.

In superalloys cast in the normal way, the individual gamma prime particles — typically less than 1 μm across — each consist of a single domain. Consequently, it has not been possible to manipulate their domain size, which is an important variable in the general theory of the mechanical behaviour of ordered alloys. This situation has recently been changed, however, by research carried out in the Japanese materials laboratory at Tohoku University, where so much important metallurgical research has been done. An important paper from this group examines the behaviour of superalloys ultrarapidly quenched from the melt[1]. At cooling rates approaching 10^6 K/s, there is not time for proper ordering and the phase consists of fine antiphase domains around 50 nm in size. The result, for a variety of phases of type Ni-Al-X, where X is Cr, Mo or Fe, and the alloys consist entirely gamma prime, is that both the strength and the ductility of the alloys are enhanced, whereas usually one of these improvements must be bought at the expense of the other. Once the quenched alloys are annealed, the domains grow (the state of order is improved) and the alloys become both weaker and less ductile. Unfortunately, Inoue *et al.* did not test the variation of the strength of the fine-domain structure with temperature; it is now important to know whether that is also subject to anomalous behaviour.

Another recent Japanese paper, from Tokyo[2], reports the extent to which third metals can be dissolved in the gamma prime phases Ni_3Al, Ni_3Ga and Ni_3Ge, and the superlattice site preferred by the third metal; these features are related to atomic size ratios and to the various bond strengths. In the present context, it is of particular interest that two of the authors of that paper previously studied the anomalous mechanical behaviour of a number of gamma prime phases (for example, Pt_3In and Pt_3Sn), finding, in general terms, that the more stable the ordered phase the more pronounced the anomalous strengthening with rise of temperature[3]. This seems consistent with the orthodox explanation of the anomaly, outlined above.

It should now be possible to answer some very interesting questions by rapid-quenching experiments on such 'model' gamma prime phases, both binary and ternary. First, does the size of the microdomains produced by quenching scale with the stability of the ordered phase? Second, how do the mechanical characteristics of these phases, particularly their anomalous behaviour, react to the introduction of imperfect order — that is, small domains? Finally, what are the consequences of the gradual healing of imperfect order by subsequent heat-treatment? In this connection, it is probable that the domain size in a ternary alloy series based, say, on Ni_3Al or on Ni_3Ga, will be a function of concentration of the third element, and if so, it will be illuminating to see how the mechanical anomaly varies across a ternary series prepared by melt-quenching.

In this way, the curious mechanical characteristics of gamma prime can perhaps be related to the mechanical behaviour of imperfectly ordered phases such as Cu_3Au or FeCo, on which a large literature exists. In general, the strength of such alloys peaks at an intermediate degree of order, but the dependence of ductility on the degree of order has not been studied. The introduction of rapid quenching, it seems, may re-open the study of the mechanical characteristics of ordered phases, which has been in the doldrums of late.

1. A. Inoue, H. Tomioka and T. Masumoto, *Metall. Trans.* **14A,** 1367 (1983).
2. S. Ochiai, Y. Oya and T. Suzuki, *Acta metall.* **32,** 289 (1984).
3. T. Suzuki and Y. Oya, *J. Mater. Sci.* **16,** 2737 (1981).

38

Order in Disorder

Nature, **335**, 493 (1988)

Since Zachariasen's seminal work[1] on the structure of oxide glasses it has been commonly accepted that such glasses are made up of a partially continuous network of SiO_4 tetrahedra, interspersed irregularly with alkali and alkaline-earth metal ions. The experimental evidence for this proposition, mostly from X-ray scattering, has built up steadily over the years — for example, from the famous study of vitreous silica by Mozzi and Warren[2]. Nevertheless, the second half of the proposition, what might be called the currant-bun model of the distribution of the metal ions, has never been established as firmly as the idea of the SiO_4 network. On page 525 of this issue of *Nature* [3], Eckersley *et al.* marshall evidence indicating that the calcium ions in a calcium silicate glass have a first-neighbour shell ordered nearly as well as the oxygens around silicon atoms.

A common experimental trick for focusing structural information on a particular species in a glass is to determine scattering profiles for X-rays of two different wavelengths, one near an absorption edge for the element in question, so that scattering amplitudes are different; a differential Fourier transformation then gives specific information about the environment of the targeted atomic species. An even better way of achieving this end is to scatter neutrons, using two successive samples containing different isotopes of the target atom, with widely different scattering cross-sections.

Eckersley *et al.* use the latter approach with the calcium in their glass, and establish that, out to a distance of about 0.5 nm (2 atomic diameters), the environment surrounding calcium ions is closely similar to that in crystalline $CaSiO_3$ (which has a composition close to that of the glass examined). Specifically, about 85 per cent of the oxygen ions nearest to a calcium atom in the glass must be disposed similarly to the corresponding ions in the crystal. A further implication of this finding, as the authors prove by careful calculations, is that calcium ions are not randomly distributed but have rather well defined, preferred separations. The venerable network model must thus be modified in the direction of enhanced order.

Although the network model is firmly entrenched for oxide glasses based on silica, it is less so for metallic glasses (for example, Ti-Cu), over which a battle — by no means entirely semantic — has raged between the adherents of the dense random-packing model introduced by Bernal and those of the microcrystallite model favoured by Gaskell[4]. For the latter model, small groups of metallic atoms form virtually crystalline unit cells which are, of course, not disposed in parallel orientation to each other. Eckersley, Gaskell *et al.* make it specifically plain that their new findings concerning the calcium silicate glass do not support the microcrystallite model for that glass.

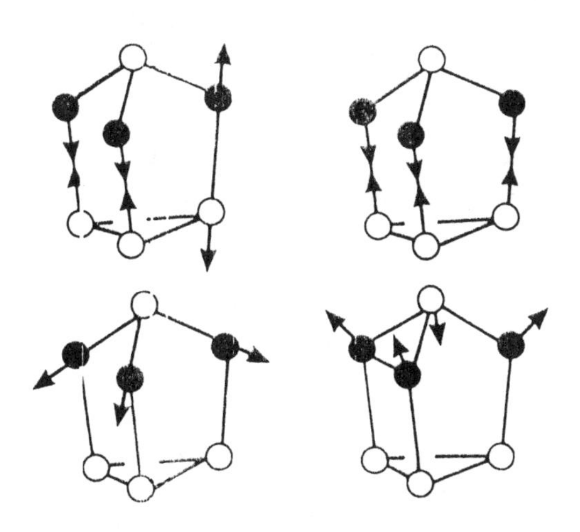

Four of the ten possible vibrational modes of the molecule P_4Se_3. The observation of these and other modes in the vibrational spectrum of P_4Se_3 glass permits Verrall and Elliott[5] to infer the presence of molecular units in this system.
(o, phosphorus, • selenium.)

A third category of glass is the molecular variety, where complete, undistorted individual molecules (monomers or poly mers) are disposed in a non-crystalline array. These have hitherto all been organic materials, which are difficult to study with precision by scattering methods because the constituent atoms scatter X-rays and photons only weakly. Verrall and Elliott in a new study[5] of a glass of composition close to P_4Se_3 establish that at least this inorganic glass is essentially molecular in its structure. Curiously, although the method they use again involves neutrons, in this case they were used to determine the vibrational density of states, not coordination shells of the material.

The use of spectroscopy, as distinct from scattering studies, to obtain information about glass structure is well established[6,7], especially for oxide and chalcogenide (sulphide, selenide and so on) glasses. What is new about Verrall and Elliott's study is the application of the method to a glass consisting, as it turns out, of well defined, distinct, inorganic molecules (see Figure). The vibrational density of the states for the glass shows remarkably sharp peaks, normally charateristic of discrete molecules, and a comparison with Raman spectral peaks for crystalline P_4Se_3 shows an excellent correspondence. The observations[5] even allow an estimate to be made of the nature of the intermolecular material when another P-Se glass, of divergent composition, is examined. Perhaps in due course a proper diffuse X-ray scattering study will be undertaken to establish just how, statistically, the molecuies are disposed with respect to each other.

The conclusions of Verrall and Elliott are particularly interesting in view of current debates about structural issues in respect of chalcogenide glasses[8]. Thus, pressure-optical experiments on glassy GeS_2 established that this glass consists of an irregular array of two-dimensional layers rather than a three-dimensional polymeric array. The new work by Verrall and Elliott adds to the structural variety of germanium-based chalcogenide glasses.

1. W.H. Zachariasen, *J. Am. Chem. Soc.* **54**, 3841 (1932).
2. R.L. Mozzi and B.E. Warren, *J. Appl. Cryst.* **2**, 164 (1969).
3. M.C. Eckersley, P.H. Gaskell, A.C. Barnes and P. Chieux, *Nature* **335**, 525 (1988).
4. P.H. Gaskell, *Phil. Mag.* **32**, 211 (1975).
5. D.J. Verrall and S.R. Elliott, *Phys. Rev. Lett.* **61**, 974 (1988).
6. J. Zarzycki, *Les Verres et l'Etat Vitreux* (Paris, Masson, 1982). [English version, *Glasses and the Vitreous State* , published by Cambridge University Press in 1991.]
7. S.R. Elliott, *Physics of Amorphous Materials* (London, Longman, 1984, and subsequent editions).
8. R. Zallen, in *Amorphous Metals and Semiconductors* (eds. P. Haasen and R.I. Jaffee) pp. 325-333 (Oxford, Pergamon, 1986).

Glasses

39

Glassy Alloys

Nature, **252**, 100 (1974)

Since times prehistoric, metallurgists have quenched iron or steel in water to harden them. Nowadays they know that the hardness is linked to the presence of a metastable crystal form, containing more carbon in solution than iron should at ambient temperature. This crystal form, martensite, has attracted immense research attention, because the genesis, hardness and breakdown of martensite are all absorbingly interesting scientific problems.

It is surprising, then, that it was not until 1960 that anyone reflected seriously upon the possibility of improving on the fastest rate of quenching then available: dropping a sliver of steel into a bucket of water might cool it at a measly rate of perhaps 1 000°C per second. It was always a fair guess that the faster one can quench an alloy, the less chance it has to convert to the equilibrium structure and the more likely it is to be fixed in some abnormal and potentially intriguing crystal structure. Yet no one seriously went into the matter until Pol Duwez at California, Institute of Technology, in 1960, recognized that the only effective way to speed up cooling rates substantially was to start from the liquid state. He conceived the idea of blasting a small molten drop of alloy by means of a gaseous shock wave against a sloping piece of copper: the technique soon acquired, over the opposition of its fastidious inventor, the onomatopaeic designation "splat-cooling". The drop is atomised and then "splatted" out into a thin liquid layer, which gives up its latent heat to the copper in a very short space of time.

We now know that splat-cooling and related techniques can generate cooling rates from 10^6 to as much as 10^9 degrees C per second. A complete new metallurgy has been created: highly supersaturated solid solutions, a whole zoo of new metastable crystal structures, and glassy alloys have all been developed. Several hundred research papers have recently been reviewed by Jones[1]: the work covers not only structures but also anomalous mechanical, electrical and magnetic properties.

Of late, there has been a sudden burst of research on the least understood of the new metallurgical species, glassy alloys. The interest stemmed primarily from the work of David Turnbull at Harvard. Turnbull is renowned for his studies on crystallization of molten metals, in particular his classic studies on the freezing of molten microspheres. It was a natural transition for him to examine the conditions for the congealing of liquid alloys into a glassy state, uncrystallized. He adopted one of Duwez's early alloys, Pd-Si, which could readily be liquid-quenched to a glassy state, and studied its crystallization on reheating (for example, Chen and Turnbull[2]). Turnbull's ideas then inspired a metallurgist, John Gilman, to more systematic studies. It had already been recognized that alloys of transition metals with metalloids were the most sluggish to crystallize and therefore aptest to form glasses. Gilman became research manager at Allied Chemical Corporation at Morristown and instituted a systematic programme of study. His colleagues, including several of Turnbull's former collaborators, examined a range of alloys based on nickel or iron or mixtures of both, with substantial additions of carbon, boron, phosphorus, silicon and aluminium. They standardised on the "roller-quenching" technique: a fine stream of molten alloy impinges on a pair of rollers, very rapidly rotating in contact, and a narrow continuous glassy ribbon emerges. These glasses typically have a glass transition temperature, T_g, in the range 650-720 K and a crystallization temperature, T_c, some tens of degrees higher. Polk and Chen[3] have recently reported on the characteristics of some of these alloys. One generalization which emerges is that the stablest glasses — that is, those with the largest value of $(T_c - T_g)$ — are those which contain the lowest metalloid content consistent with glass formation. The metalloid solute seems to exert its crystallization-inhibiting effect by a combination of simple jamming of metal atoms (because the small metalloids fill up the voids between the metal atoms) and strong bonding between solvent and solute, which inhibits displacive rearrangements. Typical compositions include $Fe_{75}P_{15}C_6Al_4$ and $Fe_{38.5}Ni_{38.5}P_{18}B_2Al_3$.

The glassy ribbons are extremely strong, with ultimate tensile strengths typically exceeding 200 000 pounds per square inch (Americans are incurably addicted to these venerable units). Several of these alloys, notably nickel-based alloys such as Ni-P-B-C-Al, when the roller-quenching rate was high enough, combined very high strength with measurable ductility (Chen and Polk[4]). This is a unique combination: normally very high strength goes with extreme brittleness, as with fine silicon or carbon fibres. The manufacturers expect to apply their ribbons to reinforcing functions, for instance in car tyres.

The physical mechanism underlying the limited ductility of the alloy glasses is also strikingly novel. Pampillo and Reimschuessel[5] have recently shown that the ribbons deform by a process akin to glide in a metallic crystal. Thin layers of alloy, inclined at 45° to the tensile axis, behave as if they were liquid and the two halves of the ribbon slide over each other. This is probably due to the incipient destruction of short-range (glassy) order and the consequential weakening of resistance to glide. Fracture initiates at several points and the cracks spread along an already defined glide plane, and eventually the cracks collide and the crystal tears apart, showing on the fracture surfaces a pattern of veins which mark the loci along which the cracks collided.

A great deal of work is now in progress on the properties, especially the mechanical behaviour, of these alloy glasses and a number of papers is expected soon in several journals.

1. Jones, *Rep. Progr. Phys.* **36**, 1425 (1973).
2. Chen and Turnbull, *Acta metall.* **17**, 1021 (1969).
3. Polk and Chen, *J. Non-Cryst. Solids* **15**, 165 (1974).
4. Chen and Polk, *J. Non-Cryst. Solids* **15**, 174 (1974).
5. Pampillo and Reimschuessel, *J. Mater. Sci.* **9**, 718 (1974).

40

Glassy Crystals

Nature, 253, 310 (1975)

In his stinging attack on Plato's political philosophy, Karl Popper[1] castigates Plato's "methodological essentialism", the delusion that to seek, by defining it, the essence of a thing can be a valid way of arriving at knowledge about the external world — including the world of social relations. Popper assures us that a true scientist always steers clear of questions which tempt him to definition-mongering, such as "What is an atom?" I am not at all sure that this is universally true: a passion for definitions sometimes provokes a scientist into a programme of experimentation which he might otherwise never have undertaken. The quest for the word provokes the deed.

These reflections are aroused by an unusual paper on the glassy state (Suga and Seki[2]). The authors, chemists at Osaka University in Japan, consider various ways of making non-crystalline solids other than by supercooling of a liquid, and ask whether it is plausible to call such solids "glasses". By itself, such a question deserves all the obloquy that Popper would direct upon it; but Suga and Seki went on to point out that the existence of a glass transition temperature is the crucial characteristic that has gradually come to be the defining criterion for glassiness. Thereupon they were moved to undertake a systematic study of glass transitions in a number of pure compounds nudged into non- crystalline states by cooling the melt, by condensing the vapour or by other more unorthodox means. In the course of these studies, performed in specially designed precision calorimeters (DTA apparatus), the authors discovered a number of anomalies: one compound might show more than one glass transition, and the concept of glassiness therefore required subdivision. It is rather as though St Thomas Aquinas had conducted microscopic observations on the creatures cavorting on his needle-point and decided that it was necessary to distinguish archangels from mere angels.

Apart from a number of simple organic compounds such as

isopentane, methanol, isopropylbenzene and cyclohexanol, the authors also examined a number of inorganic and organic high polymers, some inorganic materials (H_2O, $SnCl_2.2H_2O$) and a range of compounds that form nematic or smectic (liquid crystal) phases. Thirtyeight materials were studied.

With regard to many of the simple compounds, it turned out that ordinary glass transitions, in the form of anomalies in the specific heat, turned up irrespective of whether the non-crystalline forms were prepared by supercooling a liquid, by condensation from the vapour or by rapid precipitation from solution. This answered the original question.

But three kinds of anomalous glass transitions were also found in some materials in addition to the ordinary transitions. Some compounds (for example, cyclohexanol, 2,3-dimethylbutane) have been known for some time to possess an intermediate structure at temperatures between full crystallinity and full liquidity, the so-called 'plastic crystal' state. The molecules are positionally ordered but orientationally disordered and mobile. This makes the molecules dynamically pseudo-spherical; because of this, they crystallize in face-centred cubic or body-centred cubic structures, which are unusually prone to plastic deformation: hence the name. Suga and Seki found that by rapid cooling they were able to freeze in this anomalous state; they christened the result a *glassy crystal* ; on gradual reheating, the metastable glassy crystals passed through a 'glass transition' at which the metastable form transformed irreversibly into a fully ordered crystalline polymorph.

Another form of glassy crystal was found by supercooling of water, $SnCl_2.2H_2O$ or $SnCl_2.2D_2O$. Here the partial disorder was attributed to a freezing-in of the irregular positions of protons or deuterons; in this connection, the different transition temperatures for the last two compounds were of particular interest. There is a particularly full historical discussion of glassy ice — which, incidentally, has nothing whatever to do with the notorious polywater of a few years ago!

Finally, Suga and Seki examined a number of liquid crystals, in which there is orientational order, accompanied by positional disorder of molecules. It proved possible to 'freeze' this state by rapid cooling. The steady variation with temperature of the degree of molecular alignment, in the 'swarms' which are the liquid-crystal equivalents of ferromagnetic domains, is thereby inhibited. When the glassy liquid crystal, as the authors call this state, is heated through a transition, the molecules regain their labile ability to vary their degree of alignment. Two compounds — cyclohexene and ethanol — were found to possess two anomalous transitions each as well as a normal transition. The nature of the two anomalous glassy crystalline states is not yet known.

The authors document their findings not only by locating glass transitions but also by computing the variation of entropy with temperature: from these plots they are able to make further deductions about the nature of the change they have observed.

The concept of a glassy crystal was sometime a paradox, but now the time gives it proof. Suga and Seki's paper is an impressive achievement which should stimulate many structural studies to confirm and extend their bold but sometimes tentative interpretations. Such studies may also have a bearing on topochemical reactions (see the detailed and very instructive review by J. M. Thomas[3]). Topochemistry is concerned with the nature and kinetics of chemical reactions, including polymerization reactions, between adjacent molecules in crystals, and such processes are bound to be modified if the state of positional and orientational order of molecules can be varied and then frozen into a particular pattern. Perhaps compounds can now be found in which internal reaction mechanisms will change at glass transition temperatures. If this proves to be so, we shall need a new definition to denote the process: vitrochemistry, perhaps?

1. Karl Popper, *The Open Society and Its Enemies* , chapter 3 (London: Routledge, 1945).
2. Suga and Seki, *J. Non-Cryst. Solids* **16**, 171 (1974).
3. J.M. Thomas, *Phil. Trans. R. Soc. (Lond.)* **277**, 251 (1974).

41

Polymorphs and Amorphs

Nature, **257**, 356 (1975)

The central puzzle about glasses is this: why do some melts form glasses on slow cooling while others do not? As with all such broad questions, the golden nugget of understanding recedes further and further as hypotheses pile up. If an explanation is attempted in terms of the viscosity/temperature relation, then one wants to know what determines this relation; if in terms of critical nucleus size and diffusion rates, then again one asks how these are related to chemical composition. As more and more uncoventional glass-formers are discovered, it becomes increasingly difficult to establish what chemical or structural features they have in common. A radically novel hypothesis therefore deserves close attention.

Such an hypothesis is advanced in this issue of *Nature* by Goodman[1]. The central observation on which his hypothesis is based is: "A surprisingly large number of the systems which form glasses have as a major (sometimes the only) constituent, a material that exists in two or more polymorphic crystalline forms which differ but little in free energy", and he goes on to exemplify SiO_2, BeF_2, PbO, As_2O_3, TiO_2, Se, S and CdP_2. He then asks whether polymorphism could be a necessary condition for a material to be a glass-former. The rest of his extremely interesting and well-documented paper is devoted to exploring, admittedly in purely qualitative terms, the implications of his novel hypothesis.

He suggests that, above the glass transition temperature T_g, transient "flicker-clusters" of all possible polymorphs should coexist, as happens for instance in liquid water. The crucial point here is that none of these clusters can reach a viable size for crystal growth, because the various clusters of polymorph A would be impeded by the clusters of polymorph B, and in particular the separate A clusters could never achieve mutually epitaxial orientation relationships. As the melt cools, eventually the clusters become firmly bonded across parts of their surfaces, leaving liquid-like material in other, unbonded interstices. This stage corresponds to T_g . The role of the minor additives that assist glass formation might well be to concentrate in the residual interstices and reduce the strain resulting from differential thermal contraction of the distinct A and B clusters. (Goodman regards it as an important test of any hypothesis whether it can interpret the important role of some glass-modifier additives which have a major glass-stabilising effect in quite small concentrations.)

Goodman suggests various experimental tests of his ideas: for instance, ways are available to check on the coexistence of microcrystalline clusters and liquid-like interstitial layers. Also, recent work on the radial distribution function in vitreous silica, combined with dilatometric and high-pressure studies, have yielded evidence for the simultaneous presence of several micro-polymorphs. He goes on to point out certain categories of materials which on the basis of his hypothesis should not form glasses: there are thus the beginnings of a predictive capacity for the model.

Particularly interesting is the final section of Goodman's paper, where he suggests possible practical implications of his ideas. Thus, the solid cluster/ liquid layer model implies that near T_g , a glass might behave as an effective absorber of gases, like a zeolite, and points out that this is consistent with the known action of glass films on the surface of semiconducting devices. He suggests that carbon melts containing 'solvent catalysts' such as iron or nickel (used for diamond synthesis) might be capable of being quenched into the glassy state. Finally, he suggests that Griffith cracks, which determine the tensile strengths of glasses, might be controlled by the microstress systems and chemical heterogeneity of his postulated heterocluster arrays. He does not say it, but it might well be that the scale of the cracks (and hence strength level) would be determined by the size distribution of the clusters, and one then becomes interested in what deter mines this distribution!

One feature of many oxide glasses to which Goodman does not address himself is the observation that many (perhaps most) such glasses undergo liquid-in-liquid separation above Tg : electron microscopy and X-ray small-angle scattering both show that disperse drops of liquid separate out in a matrix of liquid of distinct chemical composition. It is not immediately clear whether Goodman's hypothesis would imply that each liquid phase needs to break down into arrays of two (or more) types of polymorphic microclusters, or whether it suffices if one phase does so. In any case, it would appear that in view of the prevalence of liquid separation, Goodman's ideas imply that there are often two distinct levels of structural heterogeneity in a stable glass.

This paper will suggest to the reader all sorts of new consequential hypotheses and various kinds of experiments to follow them up. Whether or not time fully confirms his ideas, his paper will undoubtedly prove to be of seminal importance.

1. C.H.L. Goodman, *Nature* **257**, 370 (1975).

42

Metallic Glazes

Nature, **260**, 285 (1976)

A number of alloys based on transition metals — notably nickel and iron — can be made in the form of glassy ribbons by ultra-rapid quenching of a molten jet on a rapidly spinning roller. Iron-based glassy ribbons containing chromium have the further distinguishing feature of being extremely corrosion-resistant. These materials, however, have to be in the form of thin (0.05 mm), narrow (1-2 mm) ribbons, and there is not a great deal that the most ingenious designer can do with them. Much interest therefore attaches to a report of a novel method that allows the surface of bulk material to be converted into glass without affecting the interior.

A group of metallurgists at the United Technologies Research Center in Connecticut, led by Bernard H. Kear and Edward M. Breinan, have described to the American Society for Metals how a continuous carbon dioxide laser can be used to 'glaze' the surface of nickel or cobalt-based alloys. The Center is linked with a factory that makes aircraft engines, and the implications of the choice of alloys are clear enough.

A finely focused beam from the CO_2 laser is passed over the surface of the bulk material; the surface is rapidly melted and as rapidly re-frozen as the hot spot moves on. A cooling rate exceeding a million degrees per second was claimed, during resolidification, and this suffices to generate a glassy (non-crystalline) phase at the surface. The alloy compositions used are not specified: it would be interesting to know whether they contain metalloid solutes such as P, C, B, Si, as do the alloys used to make glassy ribbons. It is particularly worth noting that the underlying crystalline grains do not act as nuclei for the resolidifying surface film.

The investigating team reports that the surface glaze is harder than the base alloy, and adds that it is "potentially more capable of resisting corrosion than the original alloys": the cautious choice of words leaves one uncertain whether the potential had been realised.

This is not the first time that a laser has been used to effect rapid quenching from the melt. Laridjani, Ramachandrarao and Cahn[1] used a stationary beam from a pulsed ruby-laser to produce metastable crystalline phases on the surface of Ag-Ge alloys, and Elliott, Gagliano and Krauss[2] used a pulsed neodymium laser to create a continuous series of solid solutions in a range of Cu-Ag alloys which are only partially miscible in equilibrium. The United Technologies metallurgists have improved on this earlier work in two ways: they examined alloys based on transition metals which, unlike copper and silver-based alloys, are apt to form glasses on melt-quenching, and they used a moving beam from a continuously acting laser to scan the surface. What is not clear from the report is whether it was feasible to make a series of overlapping passes in such a way to cover a surface entirely with a glazed layer. If this can be done, the industrial potential is very considerable; it should in principle be possible to 'coat' even the inside of a vessel with the aid of a guided moving mirror. There is also potential for the design of wholly or partially glazed bearing surfaces.

1. Laridjani, Ramachandrarao and Cahn, *J. Mater. Sci.* 7, 627 (1972).
2. Elliott, Gagliano and Krauss, *Metallurgical Trans.* 4, 2031 (1973).

43

Metallic Glasses: the Reason Why

Nature, **274** , 848 (1978)

Metallic glasses in great variety have been made by quenching appropriate liquid alloys at ultrafast rates, $10^5 - 10^8$ K s^{-1}, into the form of thin ribbons or disks. Most such glasses consist of metal/metalloid combinations but some are also made from metal-metal melts. The degree of (meta)stability of such a glass against crystallization can be measured in two ways: the critical cooling rate required to inhibit crystallization during the quench can be estimated, or the temperature at which the glass begins to crystallize on reheating can be determined. The former measure, sometimes called 'glass formability', is to be preferred, because it is related to the maximum rate of crystallization of the glass (located at a temperature well above that of incipient crystallization); it is the maximum speed at which a crystal can consume the glass that matters, not the temperature at which it happens.

A question which has exercised metallurgists and physicists ever since the first liquid-quenched metallic glass was discovered in 1959 is: what atomic or structural features determine the degree of glass formability? Three alternative models have been proposed. They are not mutually exclusive, but it is nevertheless of considerable interest to know which factor is the crucial one for any particular glass, or even for metallic glasses in general. One model, formulated by Polk in 1972, envisages simply that metallic atoms form a dense random packed array and the smaller metalloid atoms fill the voids in that array and thereby stabilise it. This model cannot be of universal validity, since it cannot account for the formability of the metal-metal glasses. A second model, originated by Chen and Clark in 1973, points to the X-ray evidence for the existence of short-range chemical order between first-nearest neighbour atom pairs in metallic glasses and melts (evidence which has continued to accumulate) and attributes easy glass formability to the enhancement of cohesion in the liquid/glass resulting from strong, chemical bonding between unlike nearest neighbours. The third model, conceived by Nagel and Tauc in 1975, attributes the high resistance of certain glasses against crystallization to purely electronic factors. This last model, intended as a critique of the other two, seemed to be riding high, but now its claim to universality has been overthrown by a new experimental study[1].

The free-electron model of S. R. Nagel and J. Tauc[2,3] is in effect a modification of Mott and Jones's classical model for the stability of certain crystalline phases such as β CuZn. Nagel and Tauc point out that a glass must have a spherical Fermi surface of radius k_F and, although no reciprocal lattice vectors exist, there is an analogous quantity which is experimentally determinable; this is the structure factor $S(Q)$, a function of reciprocal distance, closely related to the interference function and the pair distribution function. The structure factor of any glass (or liquid) always has a high and narrow first peak, at reciprocal distance q_P from the origin, and the main postulate of Nagel and Tauc is that high glass formability is favoured by the condition $2k_F = q_P$. This is equivalent to the condition that the Fermi energy E_F lies at a minimum of the density of states curve: this in turn implies that the glass reposes in a metastable minimum of electronic energy with respect to changes in composition and therefore in E_F. In their 1977 paper Nagel and Tauc explain, ingeniously, how this criterion can lead naturally to a preference for unlike nearest neighbours without the need to postulate strong interatomic bonds, though the nagging suspicion will not go away that this distinction amounts merely to semantic sleight-of-hand! They also show that a number of actual glasses in the Au-Si, Au-Ge and Co-P systems obey their criterion.

What Bauhofer and Simon have done is to make and examine glasses in two unexpected systems, Cs-O and Rb-O, and show that these glasses do not at all obey the Nagel-Tauc criterion. Though pure alkali metals are stereotypical non glass formers, Cs-O melts can be turned into glasses over the range 15-20 at.% O, and Rb-O over a narrow range near 13 at.% O (which does not incidentally coincide with the eutectic minimum at ≈ 9 at % O); the Rb-O glasses have a small admixture of crystallites. These glasses all crystallize at about 200 K, so that to make them, a sub-zero quenching medium is needed. It turns out that quenching the melt, contained in a 2mm capillary, direct into liquid nitrogen is effective, in spite of the fact that the cooling rate cannot exceed 100 K s^{-1}.

Bauhofer and Simon also measured the X-ray scattering of their glasses and thus located the first peak of $S(Q)$ in each case. They then adduce evidence that oxygen dissolved in liquid caesium removes two electrons per oxygen atom from the conduction band to form an O^{2-} ion; this also happens in crystalline Rb and Cs suboxides. They conclude that the Fermi level can be computed on this basis, and when this is done it turns out that, for both glasses, k_F coincides accurately with the *valley* between the first two peaks of the structure factor. Thus, $2k_F = q_P$ is not obeyed for these very readily formable glasses, and the authors conclude that strong though non-directional chemical bonding in the glasses is the basis of their stability. To form, for instance, crystalline caesium suboxide, large, strongly bonded $Cs_{11}O_3$ clusters have to form first and this is bound to be slow; presumably similar clusters exist in the glass. The electronic criterion of glass formability, whatever its merits eventually turn out to be, is not universal.

1. Bauhofer and Simon, *Phys. Rev. Lett.* **40**, 1730 (1978).
2. S.R. Nagel and J. Tauc, *Proc. 2nd Int. Conf. on Rapidly Quenched Metals* (ed. Grant and Giessen) p. 337, (Cambridge, MA, MIT Press, 1976).
3. S.R. Nagel and J. Tauc, *Solid State Commun.* **22**, 129 (1977).

44

Aluminium-based Glassy Alloys

Nature, **341**, 183 (1989)

A new family of glassy alloys based on aluminium, announced by three independent groups[1-5] over the past two years, is eliciting much interest. Being both strong and ductile, the alloys may succeed in technical applications where other metallic glasses have disappointingly failed. Furthermore, as described recently by Dubois and colleagues[2], these glasses do not fit familiar theories and their compositions sometimes resemble those of the remarkable quasicrystals discovered five years ago, prompting important new scientific questions.

Metallic glasses (or glassy alloys) have been studied intensively since their discovery in 1959 by Duwez and his collaborators. The term 'glass' is used in preference to 'amorphous solid' to denote the fact that Duwez made his revolutionary materials by rapid cooling from the melt: the term 'glassy alloy' is preferable to 'metallic glass' because all such glasses have to be alloys, as pure metals cannot be made glassy — these crystallize even during cooling at rates asfast as 10°C s^{-1}. Whatever the nomenclature, the examples studied during the past three decades were almost all made of transition metals, alloyed either with other transition metals or with metalloids such as boron, phosphorus and silicon[6]. Apart from a few offbeat curiosities such as calcium-magnesium glasses and transition-metal/beryllium combinations, no other types of glass have been found — until recently.

The driving forces behind the discovery of the new alloys were first, old-fashioned scientific curiosity and second, the need to find glassy alloys whose mechanical strength can be exploited. Unlike the brittle oxide glasses widely used in glass-reinforced plastics, glassy alloys are not only very strong but also often ductile, yet repeated attempts to use them in ribbons to reinforce composite materials have come to nothing, almost certainly because they have been too dense. The only exception is the family of Ti-Zr-Be glasses whose application foundered on the toxicity of beryllium.

Over the past 17 years, small volume fractions of glassy phase have been oserved in various binary aluminium alloys on rapid solidification, notably in Al-Cu, Al-Ge and Al-Cr. Inoue *et al.* [7] were the first to turn an aluminium alloy specimen completely to glass: some of their (Fe,Co,Ni)-Al-B glassy alloys had Al contents of over 50 per cent. A little later, Shingu and colleagues examined a series of Al-Fe-Si alloys[8], but these were very brittle and the investigation was dropped. Only one of their alloys, $Al_{69.6}Fe_{13.0}Si_{17.4}$ (which is close to that of the crystalline compound $Al_9Fe_2Si_2$), was completely turned to glass by melt-spinning (the solidification of a jet of molten alloy on contact with a spinning copper wheel). Note that the two auxiliary constituents are familiar from other metallic glasses.

This discovery impressed the French metallurgist, Dubois, working at Nancy, who sought to predict glass-forming ranges in aluminium alloys from structural first principles, an approach he termed the 'chemical twinning model'[9,10]. An essential part of the model is that glass formation is most likely at compositions close to those of stable crystalline compounds — a notion consistent with Gaskell's[11], that the disposition of nearest-neighbour atoms in glassy alloys is similar to that in the unit cell of a related crystalline phase. Dubois and colleagues studied[12] the short-range order in the Al-Si-Fe glass discovered by Shingu and colleagues, finding an atomic arrangement very similar to a stable phase of Al_2Cu. Two other groups, Masumoto's at Tohoku University and Shiflet's at the University of Virginia, independently began to investigate aluminium-rich glasses, in particular their composition ranges, mechanical properties and short-range structural order. There was especial interest in achieving the highest possible percentage of aluminium, as this would assure glasses of low density.

Behind Dubois's priority in this 'race', there lies an interesting twist. Dubois and Le Caer enclosed a paper in a *pli cacheté* , a sealed envelope, and deposited it with the French Academy of Sciences on 7 June 1982. It was opened two years later, at the authors' request, and a summary[1] was submitted to and published by the Academy in 1985. It emerged later[2] that the authors had joined forces with the industrial laboratory of Pechiney, a French aluminium company, and had been taking out patents since 1982. The first patent discussed a wide range of glass-forming aluminium- rich compositions with up to five components, for example, $Al_{69}Cu_{17}Fe_{10}Mo_1Si_3$. The delay in publication achieved by the popular French device of the *pli cacheté* was probably needed because further patent applications were underway: indeed, an American patent was granted in late 1987.

Dubois's team sought to compare its first glasses, based on Al-Cu-Ni, with the recently discovered quasicrystals, most of which form in

Al-rich compositions with transition metals. It turns out that those mixtures of aluminium and transition metals that are most apt to give quasicrystals are also those that require the most silicon to form glassy alloys instead (Fig. 1). The authors have, so far, pursued the matter no further, concluding in their latest paper that "aluminium glasses inspire intellectual modesty and aesthetic contemplation in equal measure". This appealing reaction would also be apposite for recent work by Shiflet and colleagues[13], who find that Al-Fe-Gd alloys which can form glasses, in their equilibrium state have no eutectic (a configuration in which the melting point is most strongly depressed) although this is the usual feature favouring glass formation.

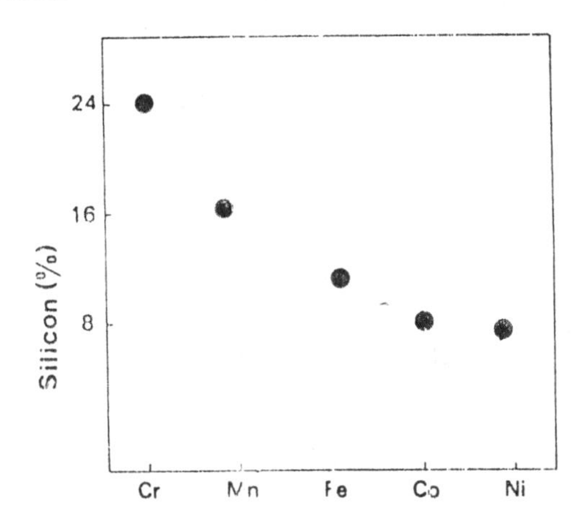

Fig. 1. Minimum concentration of silicon needed to induce glass formation in aluminium alloys[2].

Observations of rapidly quenched $Al_{86}Mn_{14}$ suggested that the product is quasi-microcrystalline, raising the problem of distinguishing such a structure from true glassiness. That distinction was achieved ingeniously by a novel calorimetric method recently designed for the purpose at Harvard[14]; it would be interesting to apply this technique to characterize some of the new Al-rich glasses.

The relationship between tendencies to form quasicrystals and glasses in aluminium, identified by the French team, is complemented by the Japanese efforts to interpret the role of atomic sizes and the strength of bonds between different atom pairs in influencing the ease of glass formation. (The claim[2] that only the French investigated the fundamentals does not really stand up to examination.) The Tohoku team, renowned for its systematic investigations of metallic glass systems, has discovered the usefulness of rare-earth metals as alloying constituents in aluminium-rich glasses; these additives are so potent that some binary and ternary alloys containing up to 90 per cent aluminium form glasses. The binary glasses (Al-Y, Al-La and Al-Ce) are difficult to make and the acceptable composition ranges are narrow, but ternary compositions (Al-Y-M and Al-La-M, with M = Fe, Co, Ni or Cu) form glasses over wide compositional ranges (Fig. 2).

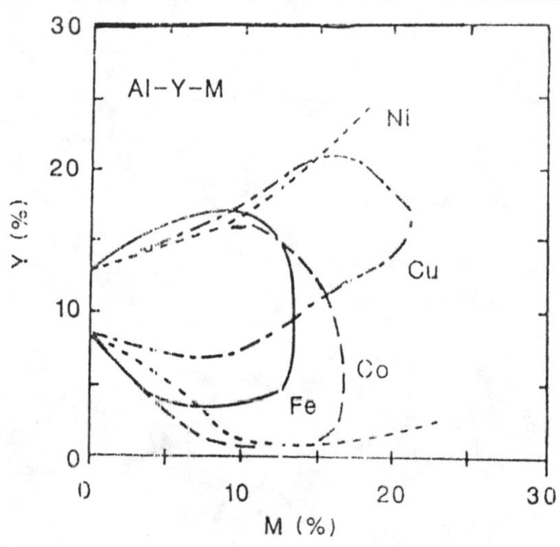

Fig. 2. Binary and ternary aluminium alloys whose compositions fall inside the loops are able to form glasses if rapidly quenched[15].

The tensile strength of these materials can be formidable — often exceeding 800 megapascal (or even 1 000 MPa in the case of Al-Y-Ni[3] and Al-Ni-Zr[16]). Ternary aluminium glasses containing transition and rare-earth metals (especially Ce, Y and Gd) studied by Shiflet and colleagues, have tensile strengths up to 950 MPa. The crucial point is that these strong glasses are also very ductile, so that they are very tough (impact-resistant). Their maximum strengths are twice those available from the best age-hardenable aircraft alloys, and could be exploited in ribbons or wires for reinforcing composites, in surface coatings or in load-bearing structures. Also, the glasses are stable against crystallization up to temperatures ranging as high as 520°C, a final point favouring their commercial exploitation.

1. J.-M. Dubois and G. Le Caër, *Compt. Rend. Acad. Sci. (Paris)* **301**, 73 (1985).
2. D. Bechet, G. Regazzoni and J.-M. Dubois, *Pour la Science* , No. 139, p. 30 (May 1989).
3. A. Inoue, K. Ohtera, A.-P. Tsai and T. Masumoto, *Jap. J. Appl. Physics* **27**, L280–L282; L479–L482; L736–L739 (1988).
4. A.-P. Tsai, A. Inoue and T. Masumoto, *Metall. Trans.* **19**A, 1369 (1988).
5. Y. He, S.J. Poon and G.J. Shiflet, *Science* **241**, 1640 (1988).
6. R.W. Cahn, in *Physical Metallurgy* , 3rd edition (eds. R.W. Cahn and P. Haasen) 1778 (Amsterdam, North-Holland, 1983).
7. A. Inoue, A. Kitamura and T. Masumoto, *J. Mater. Sci.* **16**, 1895 (1981).
8. R.O. Suzuki, Y. Komatsu, K.F. Kobayashi and P.H. Shingu, *J. Mater. Sci.* **18**, 1195 (1983).
9. J.-M. Dubois, G. Le Caër and K. Dehghan, in *Rapidly Quenched Metals* (ed. H. Warlimont) 197 (Amsterdam, Elsevier, 1985).
10. J.-M. Dubois, *J. Less-Common Metals* **145**, 309 (1988).
11. P.H. Gaskell, *J. Non-Cryst. Solids* **32**, 207 (1979).
12. J.-M. Dubois, K. Dehghan, P. Chieux and M. Laridjani, *J. Non-Cryst. Solids* **93**, 179 (1987).
13. Y. He, S.J. Poon and G.J. Shiflet, *Scripta metall.* **22**, 1813 (1988).
14. L.C. Chen and F. Spaepen, *Nature* **336**, 366 (1988).
15. A. Inoue and T. Masumoto, in *Second Supplementary Volume of the Encyclopedia of Materials Science and Engineering* (ed. R.W. Cahn) 655 (Oxford, Pergamon, 1990)
16. A.-P. Tsai. A. Inoue and T. Masumoto, *J. Mater. Sci. Lett.* **7**, 805 (1988).

Techniques of Synthesis and Investigation

45

Sputtering

Nature, **220**, 1075 (1968)

If an ion is directed at a crystal and strikes it at the right angle of incidence, it tends to follow low index crystal directions or crystal planes. This phenomenon, known as 'channelling', was discussed in this column last week[1]. When the ion is not channelled, however, a sequence of collisions takes place between it and ions of the metal lattice, and some of these sequences rotate the original momentum through more than 90° so that some metal atoms eventually emerge from the crystals. These atoms are said to be 'sputtered' and can be used to fabricate thin films. Recent papers from Professor M.W. Thompson's laboratory[2] examine the phenomenon. Because the efficiency of sputtering varies with the direction of the incident ions relative to the crystal axes, the technique is also widely used as a method of etching polycrystalline surfaces for microscopic investigation. The collision sequences may be random or they may involve 'focused collision sequences'; these last are analogous to channelling of incident ions, for they involve a series of nearly parallel mutual collisions of a row of lattice ions, focused by glancing collisions with ions on neighbouring lattice rows. These sequences are an efficient means of transporting momentum through a crystal, and sputtering could result from a series of long focused collision sequences which change direction infrequently. The recent papers describe a technique for studying the energy distribution of pulse-sputtered ions by means of an ingenious time-of-flight method, and detailed results are reported for copper and gold, bombarded with monoenergetic A^+ or Xe^+ ions. The results are compared with the predictions of various models of the sputtering process, and the conclusion is that random and focused collisions contribute roughly equally to the sputtered atoms.

Another study, by M. W. Thompson and A. D. G. Stewart, which is to appear in the January issue of the *Journal of Materials Science* [3], is devoted to the changes in the surface topology of an initially smooth metal crystal as a result of ion bombardment, which causes sputtering.

By means of scanning electron microscopy, the investigators found that a population of well formed cones were formed at the surface, their half-angles depending on the incident ion energy. The cones (Fig. 1) are a striking consequence of the anisotropy of emission directions of sputtered ions, and a quantitative theory is propounded to interpret their formation.

Fig. 1. The surface of a tin crystal following bombardment with 5 keV A^+ ions to a dose of 0.26 C/cm. (Ref. 3)

---

1. R.W. Cahn, see article 2.
2. M.W. Thompson *et al.*, *Phil. Mag.* **18**, 152, 361, 377, 415 (1968).
3. M.W. Thompson and A.D.G. Stewart, *J. Mater. Sci.* **4**, 56 (1969).

46

Analysis of Surfaces

Nature, 243, 378 (1973)

Arguably the most productive new instrument in materials science since the war has been the electron microprobe analyser. This instrument, which depends on the emission and analysis of characteristic X-rays, stimulated by a minute exploring beam of electrons impinging on a selected point on a polished specimen, has transformed the study of solidification, phase transformations, diffusion and corrosion. Its essential feature is that the chemical analysis is localized, whereas X-ray fluorescence analysis of the longer established kind, which depends on excitation by hard incident X-rays, averages the composition over several square centimetres. The wheel is, however, coming full circle with the introduction of macroscopic X-ray analysis by electron excitation, which in its latest form allows measurements of elements down to boron[1]. The use of greatly improved detectors in conjunction with non-dispersive energy analysis has steadily improved sensitivity as well as light-element detection in both microscopic and macroscopic variants of electron-excited X-ray analysis.

Electron probe microanalysis used to be regarded as a form of surface analysis, but in fact the effective depth over which composition is integrated is of the order of 1 μm. True surface analysis, by contrast, requires physical methods that explore a single surface monolayer, or ≈ 1 nm depth at most. There is now a very varied array of techniques available — the emission of photons, electrons, ions or neutral particles can be stimulated by any of these four, or by heat or an intense electric field for good measure.

More than twenty permutations of stimulation/emission exist, each with its own distinct physical theory, sensitivity, speed, depth of measurement, convenience and limitations. Electron- excited X-ray (that is, photon) emission is merely one among many. The onlooker who is concerned with using these techniques, rather than developing them, may be forgiven for being bemused. Most survey articles, of which there is no shortage, fasten on one technique to the exclusion of all others. For this reason, a new survey entitled "New Developments in the Surface Analysis of Solids" does special credit to the author, Benninghoven of the University of Cologne[2]. (The journal, *Applied Physics*, is a new English-language venture by Springer-Verlag which replaces the excellent German-language *Zeitschrift für angewandte Physik* which ceased publication at the end of 1971).

Benninghoven is a specialist in surface studies and his name is especially associated with secondary-ion mass spectrometry (SIMS), of all methods the one best suited to analyse a single surface monolayer. He describes in some detail in his survey the advantages of SIMS, (including the ability to follow chemical reactions at the surface as they progress), but devotes attention also to the various forms of secondary-electron spectroscopy (including both Auger electron (AES) and photoelectron spectroscopy, the latter usually known by its originator's designation of ESCA, i.e., electron spectroscopy for chemical analysis), and even to the very novel technique of 'soft X-ray appearance potential spectroscopy', which depends on the sudden appearance of characteristic X-ray emission lines as the energy of bombarding electrons is gradually increased. Benninghoven explains the essential physics of each technique without getting side-tracked along fascinating but inessential detours (such as the mechanism of surface-ion emission in the sputtering situation) and shows how emission of different types can be reliably distinguished, how the state of chemical combination can be assessed by some techniques but not by others, how and why the effective depth of analysis varies between techniques, which are best for analysing light elements including hydrogen, and how sensitivity varies. The examination of states of chemical combination is particularly important in many applications, but the very power of this technique (see Siegbahn[3]) has led to its use predominantly in 'bulk situations' rather than in surface studies hitherto.

Another technique, related to SIMS, is ion-scattering spectrometry (ISS). Here an exploring beam of incident low-energy ions, often He^+ or Ar^+, is scattered from the surface layer: the energy distribution of the same ions is measured after scattering, and simple calculations based on Rutherford scattering theory permit deductions as to the masses of the atoms in the surface layer. This technique is outlined in a clearly and comprehensively written survey of trace analysis techniques for solids (both for surfaces and for bulk solids) by Kane and Larrabee[4] This review can with benefit be read in association with Benninghoven's.

Overall, Benninghoven's most useful survey leaves the impression that SIMS, a technique not yet widely familiar, has particular promise as a highly sensitive and discriminating method of identifying and analysing surface species.

1. Price, *Metals and Materials* **7**, 140 (1973).
2. Benninghoven, *Applied Physics* **1**, 3 (1973).
3. Siegbahn, *Endeavour* **32**, 51 (May 1973).
4. Kane and Larrabee, *Annual Rev. Mat. Sci.* **2**, 33 (1972).

47

Magnetic Filtration

Nature, **245**, 412 (1973)

A good criterion for judging whether a piece of research is suitable for summar ising in *News and Views* is the difficulty found in choosing a heading for the possible story. This one is headed 'colloid science' [this was the category heading above the article in the original *Nature* version], but it might equally well have been 'applied physics', 'magnetism', or 'fluid dynamics', 'filtration' or 'pollution abatement'. Interdisciplinarity is all the rage, and is common enough in teams: what is much rarer, and apt to get results, is the one-man interdisciplinary team.

J. H. P. Watson is one of these monanthropic teams. He has just out lined[1] the principles of a new method of filtration. He examines the behaviour of a suspension of small paramagnetic spheres in a liquid flowing at right angles to a ferromagnetic wire when a steady magnetic field is applied parallel to the direction of fluid flow. Briefly, the particles are attracted to the wire and will adhere to it if they pass within a critical distance which is itself a function of flow rate, wire diameter, particle diameter, ferromagnetic permeability and paramagnetic susceptibility. The most important parameter or figure of merit seems to be the ratio of the fluid velocity to a 'magnetic velocity', which unfortunately is nowhere defined or explained.

Watson goes on to consider the efficiency of a filter of crumpled wire in a field, in terms of the packing fraction and the above-mentioned figure of merit, both for streamline flow and for the less tractable case of turbulent flow. Such a filter, given a sufficiently strong field and fine particles, could be highly efficient, a kind of liquid-flow analogue to the electrostatic precipitators used for smoke. Watson considers the special case of sewage. Here it would be necessary to use paramagnetic powder, such as haematite, to provide "nuclei for activated-sludge flows", and then to extract the powder from the filtrate and recycle it. Watson claims that such a filter would be 100 times more efficient than a sand filter of the same area — perhaps a rather modest factor of improvement in view of the probable capital and process costs, but interesting nonetheless. Watson suggests that a superconducting magnet would be needed to attain a sufficiently high field (no field strength is quoted anywhere) and that a 12-foot diameter filter could process 2.5×10^7 gallons a day and serve a city of 0.5 million. The economics of this proposal would be intriguing in the extreme — it is hard to see whether it would be an environmentalist's Concorde or an ultra cost-effective vehicle for recycling haematite, liquid helium and water.

The particular application proposed may provoke a measure of scepticism and the theoretical treatment, as published, may raise a number of unanswered questions, but the filtration technique is one with most interesting potential, especially if the theory can be experimentally confirmed.

1. J.H.P. Watson, *J. Appl. Phys.* **44**, 4209 (1973).

48

Powder Metallurgy: New Techniques, New Applications

Nature, **276**, 209 (1978)

Ceramic objects — crockery, fine porcelain, furnace tubes, insulators — are made by starting from a powder, binding it (with water or adhesive), sometimes pressing the mass and then firing. The bulk of the powder never melts (though part of it may); by a series of physical processes including bulk and interface diffusion, the particles solid-weld together, or 'sinter', and the pores between them close up. There is really no alternative, because most ceramics are too refractory to melt and, being brittle, the powder particles cannot be squeezed together.

The idea of adapting sintering methods to metals seemed eccentric when it was first tried in Austria early this century, and the early German term 'Metallkeramik' showed the origin of powder metallurgy as a variant of ceramic procedures. From the beginning, the attraction of powder metallurgy has been economy : a fairly complex shape can be made quickly and cheaply, without the problem of solute segregation inseparable from all forms of casting. In some special instances, powder metallurgy can create a product that cannot be matched by traditional means: the manufacture of long-lasting incandescent lamp filaments is the prime instance. (The progressive improvement of electric lamps, largely a matter of metallurgical ingenuity, is a fascinating chapter in the history of metallurgy). Apart from such special cases, powder metallurgy has generally been resticted to unsophisticated products, subject to modest demands. All this is in the throes of change, and powder-metallurgical methods are beginning to be applied to demanding functions such as gas turbine disks and highly stressed airframe components. The reason is to be found largely in two very recent technical developments — rapid solidification processing and hot isostatic pressing. Rapid solidification processing (RSP) is an American industrial development stemming directly from the techniques of splat-quenching of alloy ribbons from the melt. An alloy melt is atomised and sprayed through a helium atmospher and the fine droplets, usually in the range 25-100 μm diameter, are frozen very rapidly; the exact cooling rate is subject to many experimental variables, but an average rate of 10^6 deg C s^{-1} is quite accessible. The particles are then pressed and sintered and may subsequently be extruded or wrought in other ways. This technique of making final products has the advantages that macrosegregation is entirely eliminated, large amounts of solute can he held in metastable equilibrium which makes possible new levels of precipitation-hardening, and the consolidated product can have very fine grain sizes which in turn leads to great ductility at ambient temperature (even to superplastic behaviour.

The method is being applied to superalloys for jet engine parts, titanium alloys for airframe components and to aluminium alloys. It has been difficult to discover much about these developments, many of them shrouded in industrial secrecy, but publication has now begun. Much information about both powder quenching and consolidation is to be found in the proceedings of a conference on Rapid Solidification Processing[1].

Hot isostatic pressing (HIP) is a new technique, pioneered in Sweden, with even wider potential. A chamber (which may be a fair part of a cubic metre) is subjected to high temperature — it can be well above 1 000°C — and high uniform (isostatic) pressure — typically 1 kbar — transmitted by argon. The maintenance of such temperatures and pressures in such large volumes represents a technological achievement of a high order, with correspondingly high demands on safety precautions: in the early days of HIP, an operator was suffocated in a sea of escaped argon.

Specimens for HIP consist of loose powder contained in a sealed evacuated can of metal foil or in a shaped die of glass or steel, and the process converts them in one operation into pore-free solids. Ordinary sintering of metals, without pressure, will not reduce porosity below 3-4% and removal of that small porosity leads to a large improvement in strength, ductility, toughness and fatigue resistance at room temperature. A related function of HIP is its use for 'rejuvenation' of turbine blades. Creep of a cast blade in service generates porosity at grain boundaries, which limits creep life; HIP can entirely remove this porosity and thus effectively confer on the blade a new lease of life. This method is just beginning to be used in practice. A further development, in America, is the 'near-net shape' process. A shaped thick die of glass or steel is used to contain the powder: at high temperature and pressure, the die behaves as a quasi-Newtonian fluid and transmits hydrostatic pressure. Complex airframe components can he made from powder — which may have been made by RSP — directly as 100% dense shapes which require only a minimum of machining.

The combination of RSP and HIPping (the neologistic verb is firmly entrenched in America) offers, for the first time in many years, some prospect of enhancing the service temperature of turbine blades. This is because the problem of macrosegregation in castings has strictly limited the permissible alloy content of superalloys; RSP has removed this limitation and a new range of alloy compositions is now accessible. The main problem is that the fine grain size produced by RSP, desirable for applications at ambient temperature, is quite unacceptable for turbine blades — which must have very coarse grains. There are indications that it may be possible to achieve such grains by recrystallization after RSP followed by HIP, without loss of the precipitation-hardened microstructure that confers hot strength. In this connection, current theoretical work aimed at identifying the plastic processes that enable pore closure during HIPping may prove important, because the feasibility of recrystallization is closely linked to the amount and nature of plastic deformation previously undergone by the metal. Activity in this field is very intense.

Yet another potentially very valuable use for HIP has just been proposed by Engel and Hübner of Erlangen, Germany[2]. They HIPped, at 1 kbar and 1 260–1 360°C, samples of cemented carbide (tungsten carbide particles cemented by films of metallic cobalt, made by conventional sintering combined with partial melting). These very brittle materials are used as tips for lathe tools subject to exceptioally severe demands. It was found that, after HIPping, all residual pores were completely eliminated, being filled by cobalt, which is itself not brittle. It appears that under HIP conditions the cobalt binder can flow freely between the carbide grains. Sophisticated methods of fracture testing were employed to examine the effect of HIPping on the resistance to brittle fracture, and it was found by statistical (Weibull) analysis that the mean fracture strength could he doubled, with a concomitant reduction in scatter of strength values. A critical comparison of observed flaw sizes with those computed from measured fracture strengths proved that the specimens were behaving as ideal brittle, flaw-limited solids both before and after HIPping. This very careful study indicates that HIP of cemented carbides should greatly increase the performance and life of lathe tools; moreover, this should make economic sense as a production treatment, since cemented carbides are materials of high specific value.

1. R. Mehrabian, B.H. Kear and M. Cohen (Eds.), *Rapid Solidification Processing —Principles and Technologies* . (Baton Rouge, Claitor's Publishing Division, 1978).

2. U. Engel and H. Hübner, *J. Mater. Sci.* **13**, 2003 (1978).

49

Lord Kelvin Updated

Angewandte Chemie (*Advanced Materials* Supplement) **100**, 1031 (1988)

William Thompson, better known to science as Lord Kelvin, demonstrated theoretically as long ago as 1853 that the rate of temperature change in a cyclically loaded body is directly related to the rate of change of the sum of the principal stresses, if conditions are adiabatic. If a specimen is stressed inhomogeneously, in cyclic fashion, then the distribution of stresses is expressed by a distribution of mean ΔT (relative to the undisturbed temperature of the sample). The ΔT oscillate with the same frequency as the applied cyclic stress. (An increasing tensile stress gives a temperature fall, an increasing compressive stress, a temperature rise.) — Kelvin's principle can be thought of as the thermoelastic analogue of Faraday's principle of electromagnetic induction, established at roughly the same time.

Kelvin's principle has been exploited for some years now to map the distribution of stresses in an inhomogeneously stressed thin flat body — for instance, a sheet containing a hole or a notch — by measuring the time-averaged temperature change at each point by means of a scanning infrared emission camera: the camera is electronically controlled to read ΔT at each test point in synchrony with the frequency of the applied stress. The strategy has been used for non-destructive examination (NDE) of an object's resistance to fatigue damage. Here the cyclic stress modifies the state of the material rather than being used as as a mere probe to examine stress distributions. Incipient fracture, or delamination of a fibre-reinforced composite, leads to enhanced local temperature changes, and such damage sites show up in *vibrothermography*, as the technique is sometimes called, long before they can be detected by any other NDE method[1]. The method has been found especially favourable for NDE of fibre-reinforced composites.

However, vibrothermography can do more than provide NDE: given sufficiently sensitive equipment and appropriate theory, the distribution of stresses which would result from an applied steady load can be determined by a ΔT scan with a low-frequency cyclic load which itself is far too small to damage the specimen in any way. A new instrument based on a design by SIRA (the Scientific Instrument Research Association) in England, named SPATE (for Stress Pattern Analysis by measurement of Thermal Emission) can be used in this way. (Spatial resolution is 0.25 mm^2 and temperature resolution, 0.001°C). The infrared signal at each point the specimen at the peak of the stress cycle is averaged over many cycles to achieve an accurate estimate of the stress distribution.

Recent work[2-4] at the Aeronautical Research Laboratory in Melbourne, Australia, has now advanced the technique to new and unexpected level of subtlety. Machin *et al.* [2] found experimentally that, contrary to Kelvin's theory, not only the cyclic stress amplitude but also the mean stress affect the output of the SPATE instrument. Sustained attempts to develop the theory to account for this finding led to an expression[3] in which ΔT is linearly related to the mean stress when measured at the primary frequency, but is modified by a second term with a coefficient proportional to the square of the stress range when measured at the second harmonic (twice the primary frequency). This new theoretical relationship was obtained by taking into account the variation of Young's modulus with temperature, something that Kelvin omitted to do. This new analysis offers the prospect of using vibrothermography to separate out oscillating stress amplitudes (which mimic the distribution of stress which would be created by an externally applied load) from a fixed, *residual* stress distribution which exists in the absence of any external load.

In another recent paper[4] the Australian team has demonstrated that such separation is indeed feasible. For this purpose, the electronic processor which filters out non-primary frequencies has to be by-passed so that any component of thermal response at twice the primary frequency can be separated out. Wong and his coworkers[4] prepared an aluminum alloy plate which was plastically loaded to create a predetermined distribution of residual stresses across the plate width. These stresses were also measured with an array of strain gauges cemented to the plate before the plastic loading. The plate was then cyclically loaded, well inside the elastic range, with a non-zero mean load. Use of SPATE with the electronic bypass referred to then allowed the ΔT distribution at each of the two key frequencies to be determined, and from this the residual stress distribution could be calculated. (This was straightforward because the cyclic stress distribution was uniform across the plate). The figure, taken from reference 4, compares the stress distribution obtained in this way, with error bars, with that directly determined by means of the array of strain gauges (solid line). The agreement is very satisfactory.

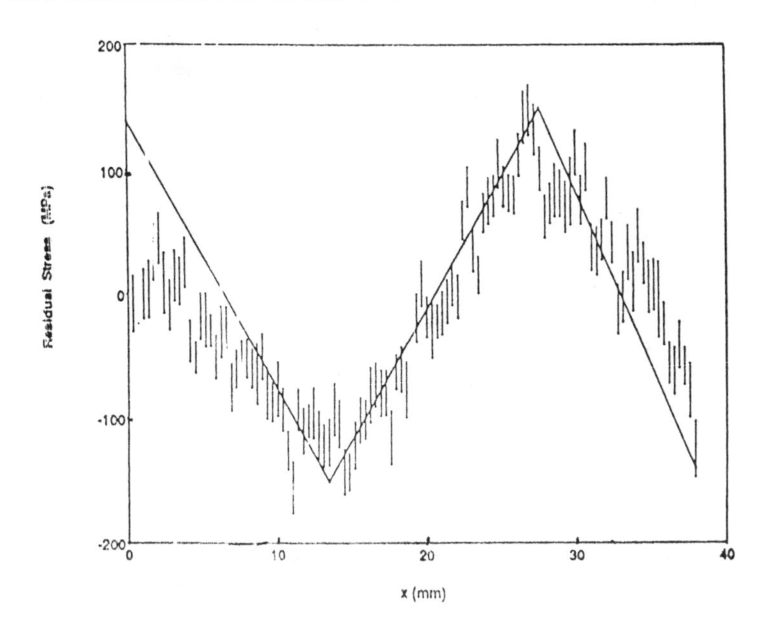

Figure. Residual stress profile for a four-point straightened specimen[4].

A few points are not quite clear in the published paper. The ordinate of the graph is simply labelled "residual stress": presumably the sum of the principal stresses is meant. Also, it is not clear whether the use of a non-zero mean load in the cyclic loading regime is essential in order to be able to determine the residual stresses; presumably a non-zero mean load helps in separating out the weak double frequency term from the background noise.

The authors point out that their demonstration was conducted under artificially simple circumstances: the simple rectangular shape of the sample, without holes or other disturbances, meant that the cyclic stress was uniform throughout the specimen. (However, the technique, in this simplified form, could be applied to difficult cases such as that of a glass sheet; the only extant non-destructive method of estimating residual stresses involves X-ray diffraction and this cannot be used with a non-crystalline material.) — With more realistic specimens of more complex shape, the cyclic stress as well as the residual stress will vary from point to point. To separate out such variables, the sensitivity of

SPATE will need to be greatly enhanced, and microstructural feature, such as grain boundaries and preferred crystallographic orientation of the grains, would also need to be taken into account in these interpretations.

1. E.G. Henneke III in M.B. Bever (Ed.), *Encyclopedia of Materials Science and Engineering* , p. 5245 (Oxford, Pergamon Press, 1986).
2. A.S. Machin, J.G. Sparrow and M.G. Stimson, *Strain* **23**, 27 (1987).
3. A.K. Wong, R. Jones and J.G. Sparrow, *J. Phys. Chem. Solids* **48**, 749 (1988).
4. A.K. Wong, S.A. Dunn and J.G. Sparrow, *Nature* **332**, 613 (1988).

50

In vino veritas

Nature, **338**, 708 (1989)

Grape juice is often 'enriched' with additional sugar, in the hope of enhancing the quality of the wine resulting after fermentation. The European Community (EC), in the shape of the Council of Ministers, occasionally debates whether or not to prohibit the practice, or to limit the amount and type of sugar that may be added, and new regulations will eventually be promulgated. However, such regulations would be pointless without a reliable method to detect the added sugar after fermentation is complete. Such a method now exists, and the EC recently approved it as a standard Community procedure.

Jean Antoine de Chaptal (1756-1832) was an agile French chemist who managed to keep his head while an eminent contemporary, Lavoisier, lost his to the post-revolutionary Terror. A landowner, Chaptal introduced the cultivation of sugar beet, the British blockade having prevented the normal importation of cane sugar. Having succeeded in this, he proposed that after summers with too little sun, grape juice be fortified with beet sugar so that it fermented as it should. Ever since, the process of fortifying wine in this way has been known among the French as *chaptalisation*. The *Supplement* to the *Oxford English Dictionary* defines chaptalization more broadly: "to convert or improve the must, in wine-making, by neutralizing an excess of acid or adding sugar". In the EC's Council of Agriculture Ministers, the French and the Germans, dwellers in cool climes, are inclined to refer to it informally as *enrichment* ; the British, more neutral in the matter, call it *sugaring* ; while the Italians and Greeks, whose hot climates ensure that grapes are never short of sugar, complain of *adulteration* .

Whichever word is preferred, the Commission has concluded that recent developments due to Gérard Martin's team at the University of Nantes will allow chaptalization to be identified *ex post facto* . Indeed, the French authorities, ever concerned to protect the reputation of French wines, some years ago offered a prize of 1 million francs for an effective method of detection, and Martin was the winner. Martin's most complete account of his technique was published last year[1].

Characterization of wines and distilled liquors has for a long time used one of two approaches: either sensory analysis, which is an impressive name for tasting and sniffing (see illustration), with a top-dressing of multivariate statistical analysis and psychology... or capillary gas chromatography, which identifies the numerous esters and other constituents which are responsible for flavour. These approaches are critically discussed in a recent book[2] and clearly have serious weaknesses — not least, the difficulty arising from the fact that the population of esters, and so on, is constantly changing as a wine matures. In the same book, Williams[3] in particular examines the problems involved in tracking down a wine to its region of origin and in testing for chaptalization, and points to the special strength of the method introduced by Martin, 'site-specific natural isotope fractionation" (SNIF).

Sentir le bouchon : olfactory sensory analysis will soon be displaced by SNIF-NMR[1]. (From*Monseigneur le Vin* , Etablissements Nicolas, Paris, 1927.)

The analysis of isotope ratios has, of course, become steadily more prevalent in geochemistry, environmental sciences and biochemistry; as practised in these disciplines, the normal tool is the mass spectrometer and isotope ratios refer to molecules or crystals as a whole. Attempts have been made to use the $^{13}C/^{12}C$ ratio to distinguish between sugars from different sources (grapes, cane, beet), as different plants follow different photosynthetic cycles. It is possible to distinguish with some confidence between wines fortified with cane sugar and unfortified ones[3], but it seems that beet sugar poses an insoluble problem with this technique (though the adulteration with beet sugar of rum, made from cane sugar, is detectable).

The drawback of the traditional method of isotope analysis is that only the overall isotope ratio in a molecule can be measured. Martin's method overcomes this limitation by measuring the proportion between hydrogen and deuterium in a specific site within an organic molecule by nuclear magnetic resonance. This type of measurement can be used to 'fingerprint', in particular, alcohols fermented from different sugar precursors. Ethyl alcohol has two distinct carbon-bonded hydrogen sites, the methyl and methylene sites, and by measuring D/H ratios for these sites separately, as well as for the water constituent of wines, Martin and co-workers can pinpoint the botanical and (to a somewhat lesser degree of reliability) the geographical origin of the alcohol (see figure).

Martin and his wife, Maryvonne, first proposed[4] the SNIF-NMR method in 1981 and gave a much more detailed account[5] in 1986. In the meantime their son, Gilles, set up a company, Eurofins, in Nantes to exploit the patented technique, and the general assembly of the *Office International de la Vigne et du Vin*, a long-established intergovernmental body with 33 member states, in January 1987 after 4 years of tests approved SNIF-NMR as an offically recognized method of "displaying the chaptalization of wine and the EC followed suit the following year.

In the SNIF-NMR method, the wine is distilled to obtain almost pure alcohol; this is necessary to enhance the precision of the NMR measurement, which depends on the alcohol concentration in the NMR sample. Recent research has established that, in general, results are more consistent if measurements are made only on the alcohol resulting from fermentation, not on the original must or grape-sugar solution. In fact, a recent (April 1988) resolution of the *Office International* has specified that only alcohol analyses be used for its purposes. The D/H ratios, which can generally be determined to a precision of 0.2 parts per million from the relative intensities of NMR lines, each identified with a distinct deuterated 'isotopomer', are calibrated against internationally standardized water samples[6], against standard tetramethylurea and against a collection of sealed standardized alcohols fermented from different precursors, provided by the Community's Bureau of References. Of course, the total weighted D/H ratio for an alcohol molecule, taking into account all sites, should match the D/H ratio obtained by mass spectrometry. According to Martin*et al.* [1] there is indeed quite good agreement but with a systematic deviation of a few parts per million, which has recently been traced to differences in the deuterium ratio of the mutually exchanging hydroxyl sites of water and ethanol (G. J. Martin, personal communication).

Over several years, Martin's team analysed many certified wines from France and other countries represented by the *Office International*, seeking correlations of the D/H ratios (for the methyl and methylene sites in alcohol and for water) with factors such as grape variety, region of origin (and hence climate) and of course chaptalization. The statistical method used is canonical variate analysis, aided by a specialized computer diagnosis system, ISOLOG[7]. Different weighted combinations of experimental variables (here, D/H ratios) are systematically compared to find the one best correlated with a particular external factor. The figure shows the results of the test of the distinguishability of groups of wines from the south of France, Switzerland and Tunisia, and the statistical separation is clear enough. The method can often distinguish between

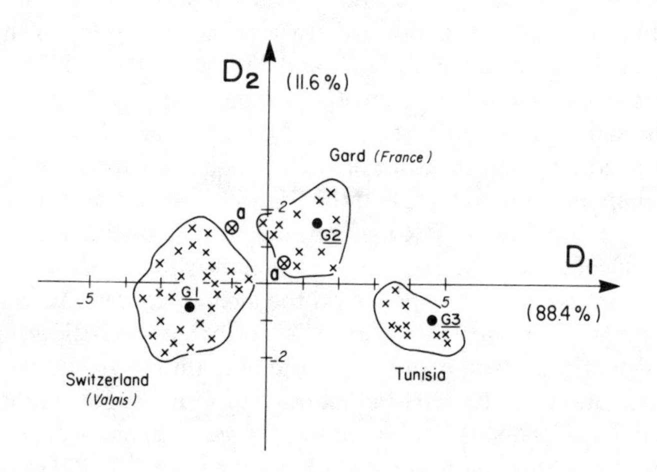

Canonical variate analysis of SNIF characteristics of several groups of wine. D_1 and D_2 are optimized weighted combinations of the various D/H ratios. (From ref. 1.)

wines made from different grape varieties. The D/H ratios seem to vary according to the sugar concentration in the various grapes, which itself is a function of geography, grape variety and the weather preceding a vintage.

As things stand, Community regulations permit enrichment of grape must only with other (more concentrated) grape must, natural or rectified, except in those countries having a long tradition of adding beet sugar, which are free to follow their national rules. The Council of Ministers some time ago instructed the Commission to report to it on current enrichment practices; this report is due soon, and then the Council will have to try to reach agreement on future regulations for Europe as a whole. It undoubtedly is an extremely sensitive political issue.

1. G.J. Martin *et al.*, *J. Agric. & Food Chem.* **36**, 316 (1988).
2. G.G. Birch and M.G. Lindley (Eds.), *Alcoholic Beverages* (London, Elsevier, 1985).
3. A.A. Williams, in ref. 2, p. 129.
4. G.J. Martin and M. Martin, *Tetrahedron Lett.* **22**, 3525 (1981).
5. G.J. Martin *et al.*, *J. Am. Chem. Soc.* **108**, 5116 (1986).
6. R. Gonfiantini, *Nature* **271**, 534 (1978).
7. G. G. Martin, F.J.C. Pelissolo and G.J. Martin, *Computer Enhanced Spectroscopy* **3**, 147 (1986).

51

Self-Propagating High-Temperature Synthesis

Advanced Materials **2**, 314 (1990)

As a schoolboy I learnt that tram (streetcar) rails in the city streets were welded together using the 'Thermit' process. A mixture of ferric oxide and aluminum powder packed in the gap between adjacent rails is ignited by an applied localized burst of heat and a self-sustaining reaction, generating metallic iron and alumina, then 'takes off'. The essence of the matter is the greater thermodynamic stability of alumina compared to that of ferric oxide: the reaction is a form of combustion. The term 'Thermit' gradually fell out of use, together with trams (though these have staged a comeback in some countries), but the word was still in use to denote the whole family of self-sustaining reactions, as recently as 1959, in a paper devoted to the synthesis of cermets[1].

The underlying mathematics of thermit-type reactions was published as early as 1953[2]. However, these reactions only began to be seriously examined through the enthusiasm of A. G. Merzhanov, head of the Institute of Chemical Physics of the USSR Academy of Sciences in Chernogolovka, near Moscow. He and his coworkers took out a patent in 1971 to cover the combustion synthesis, from the constituent elements in a sealed 'bomb', of a number of refractory carbides, nitrides, borides and other ceramics.

Merzhanov's 1972 paper[3] is taken as marking the beginning of systematic research on the process, named *self-propagating high-temperature synthesis* (SHS) by its inventors, who claimed as benefits greater speed, lower energy cost and greater homogeneity. Within a few years, some 30 Soviet organizations had become involved and the process has recently attracted a special panegyric from Mikhail Gorbachov! While I was writing this article it came to my attention that a course on SHS was being organized at the Gorham Advanced Materials Institute in the USA (for the end of March) at which Dr. Alexander Merzhanov was to be one of the key speakers. Several other Russians were scheduled to contribute, indicating that we are at last beginning to take advantage of Soviet expertise. Long may this cooperation continue.

The extensive Russian-language literature has recently been translated and collated by Frankhauser *et al.*[4]. Later, the technique attracted attention in Japan, the USA and elsewhere, though the apogee of American activity appears to have been brief (1984-86). The history of international research on SHS, to date, has been summarized in an excellent overview by Munir[5]. Even more recent, rather complete overviews have been written by Yi and Moore[6] and by Munir and Anselmi-Tamburini[7].

In the West, much attention has been devoted to two groups of materials — refractory carbides and borides of titanium, and intermetallic compounds. Understanding of the SHS of titanium ceramics has recently become much more quantitative, following on the detailed analysis of TiC_x synthesis by Holt and Munir[8]. Basing their efforts on earlier Russian work, they analyzed the nature (steady or oscillating) of the combustion and the (adiabatic) combustion temperature as a function of the composition of the reacting mixture; they also examined the effect of various proportions of a non-reacting diluent on the progress of the reaction, a procedure which can markedly improve the controllability of an SHS reaction. Such calculations have been taken to an even more sophisticated level, using a computer code, by three Indian metallurgists in a very recent paper[9]. These calculations refer to the SHS of TiB_2, by any of three synthesis routes: $Ti + 2\,B \rightarrow TiB_2$; $3Ti + B_4C \rightarrow TiC + 2TiB_2$; $3\,TiO_2 + 3\,B_2O_3 + 10\,Al \rightarrow 3\,TiB_2 + 5\,Al_2O_3$. (The third reaction is much cheaper than the others). In addition to the role of TiB_2 or TiC used as a non-reacting diluent, the influence of the initial temperature of the mixture (before the reaction is initiated by localized heating) on the resultant adiabatic combustion temperature and on the molten fraction of product was computed.

The molten fraction is an important parameter in defining the quality of the end-product, because a major problem associated with SHS is the frequently high porosity of the product, due to gas evolution caused by retained moisture, volume change during the reaction, etc.[10]. One way of reducing this problem is to apply pressure during and immediately after SHS[11] and the Indian authors pointed out[9] that if all or a substantial fraction of the product remains liquid during pressurization, porosity should be greatly reduced. It is therefore important to know under what circumstances this can be achieved.

Another method which has been tried recently in order to overcome the porosity problem is to use shock waves to initiate the SHS itself (see citation in ref. [11]). It turns out, however, that shock waves applied during the reaction are not effective, but immediately after the reaction they can be effective (presumably the more so if the product is still liquid — but Holt does not seem to have examined this aspect). Russian metallurgists have much experience of explosive densification (see a recent authoritative survey by Roman and Gorobtsov[12]), and they have done most on SHS involving shock waves as an integral part of the reaction.

Recently, much effort has been devoted to the SHS of structural intermetallics, notably Ni_3Al[13-15] and TiNi.[16]. One important motive is to bypass the intractable problem of obtaining prealloyed powders for sintering, which are hardly ever commercially obtainable and extremely expensive to make by atomization. One objective can be to make powders by a cheap and rapid process, an alternative one is to bypass conventional sintering and make solid objects directly. (A recent authoritative American overview[17] has laid down the desirability of "shape-limited synthesis", in which one or more reactants are formed in the final shape desired). The studies of Ni_3Al, a major candidate for structural applications[18], show that the density depends sensitively on stoichiometry and even more on the size, and size ratio, of the reactant powders. Therefore, medium-coarse Ni powder should be mixed with much finer Al powder for best results, because in this way the minor constituent can still envelop particles of the major constituent. The TiNi study[16] has broken new ground by studying in detail the effect of heating rate (up to 1500 K/min). When this is increased, so is the spontaneous ignition temperature and this in turn affects the adiabatic combustion temperature and the molten fraction. They also studied interference from undesired TiO_2 formation in detail and showed that at a critical heating rate, this interference could be almost eliminated.

There has also been progress in exploiting SHS to create alloys dispersion-hardened by fine intermetallic or ceramic particles. The proprietary XD process developed by the Martin-Marietta Corporation in Baltimore depends on the principle of mixing powders of the constituent elements of a ceramic phase with a metal or alloy powder[19] and presumably (details are jealously guarded!) the resulting fine dispersion (e.g., of TiB_2 in TiAl[20]) is the product of an SHS. Again, McLean[21] has shown how a concentrated dispersion of particles of an Al-Fe intermetallic phase can be produced in an aluminum matrix by exploiting SHS between elemental powders, with very careful control of powder particle size, heating rate and sintering temperature.

SHS is turning up in unexpected situations too. Schaffer> and McCormick[22] have found a discontinuous temperature rise during mechanical alloying (alloying by intensive ball- milling) of a mixture of Ca + CuO, and Bordeau and Yavari[23] have found clear evidence of SHS during the interdiffusion of a multilayer of two metals with a strong negative heat of mixing. Thus. an Al-Zr multilayer starts toreact explosively close to the melting temperature of aluminum, the lower-melting component (Figure), because the rate of reaction when one component is molten, and solid-state diffusion is no longer rate-limiting, becomes fast enough to overwhelm the rate of heat dissipation.

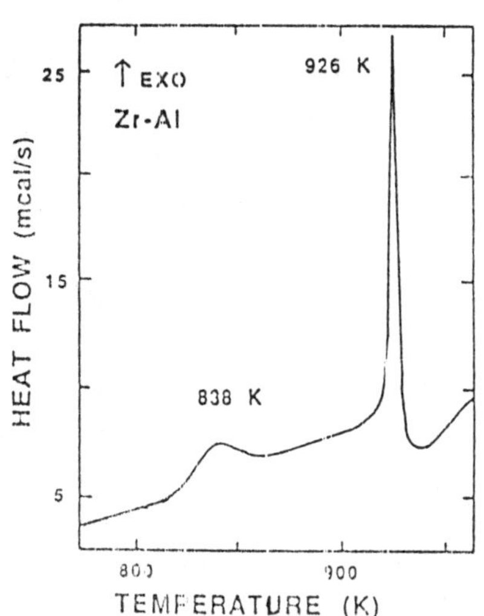

Figure. Differential scanning calorimeter thermogram of an Al-Zr multilayer. The small broad peak is due to mixing by solid-state diffusion; the sharp peak near the melting temperature of Al (933K) is due to an SHS [23]. (The Al contains a little Zr because of prior diffusion).

Bordeaux and Yavari point out the potential of such multilayers as brazing foils. Indeed, SHS has already been exploited to join dissimilar materials, by Miyamoto *et al* . [24] who joined molybdenum to TiB_2 or TiC by placing a Ti+B or a Ti+C powder mixture between Mo discs, surrounding the assembly by an insulated carbon heater and initiating an SHS reaction by means of a current pulse passed through the heater.

Clearly, self-propagating high-temperature reactions are coming now to be properly understood and are just approaching the point of widespread application.

1. J.D. Walton Jr. and N.E. Poulens, *J. Am. Ceram. Soc.* **42**, 40 (1959).

2. F. Booth, *Trans. Faraday Soc.* **49**, 272 (1953).

3. A.G. Merzhanov and I.P. Borovinskaya, *Dokl. Akad. Nauk SSSR* **204**, 366 (1972).

4. W.L. Frankhauser, K.W. Bradley, M.C. Kieszek and S.T. Sulli-

van, *Gasless Combustion Synthesis of Refractory Compounds.* (Park Ridge, N.J., USA: Noyes. 1985).

5. Z.A. Munir, *Ceram. Bull.* **67**, 342 (1988).
6. H.C. Yi and J.J. Moore, *J. Mater. Sci.* **25**, 1159 (1990).
7. Z. A. Munir and U. Anselmi-Tamburini, *Mater. Sci. Reports* **4** (1989).
8. J.B. Holt and Z.A. Munir, *J. Mater. Sci.* **21**, 251 (1986).
9. J. Subrahmanyan, M. Vijaykumar and S. Ranganath, *Metals, Materials and Processes* **1**, 105 (1989). [This is a new journal published by Meshap Science Publishers, PO Box 8319, Bombay, India.]
10. R.W. Rice and W.J. McDonough, *J. Amer. Ceram. Soc.* **68**, C-122 (1985).
11. J.B. Holt, *Mater. Research Soc. Bull.* **12**/7, 60 (1987).
12. O.V. Roman and V.G. Gorobtsov, in *Powder Metallurgy: Recent Advances* (V.S. Arunachalam and O.V. Roman, Eds.), p. 83 (New Delhi, Oxford and IBH Publishing Co., 1989).
13. K.A. Philpot , Z.A. Munir and J.B. Holt, *J. Mater. Sci.* **22**, 159 (1987).
14. A. Bose, B. Moore, R,M, German and N.S. Stoloff, *Journal of Metals* **40**/9, 14 (1988).
15. R.M. German, A. Bose and N.S. Stoloff, *Mater. Reseacrh Soc. Symposium Proceedings* **133**, 403 (1989).
16. H.C. Yi and J.J. Moore, *J. Mater. Sci.* **24**, 3449, 3456 (1989).
17. *Materials Science and Engineering for the 1990s* , p. 221 (Washington, DC, National Academy Press, 1989).
18. R.W. Cahn, *Metals, Materials and Processes* **1**, 1 (1989).
19. A.R.C. Westwood, *Metall. Trans.* **19A**, 749 (1988).
20. L. Christodoulou, P.A. Parrish and C.R. Crowe, *Mater. Research Soc. Symposium Proceedings* **120**, 29 (1989).
21. M.S. McLean, in *Horizons of Powder Metallurgy* (W.A. Kaysser and W.J. Huppmann, Eds.), Vol. 2, p. 957 (Freiburg, Germany, Verlag Schmid, 1986).
22. G.B. Schaffer and G.B. McCormick, *Scripta metall.* **23**, 835 (1989).
23. F. Bordeaux and A.R. Yavari, *J. Mater. Research* **5**, 1656 (1990).
24. Y. Miyamoto, T. Nakamoto, M. Koizumi and O. Yamada, *J. Mater. Research* **1**, 7 (1986).

Materials Aspects of Energy

52

Looking at the Hydrogen Economy

Nature, **243**, 184 (1973)

The sudden recognition that an energy crisis is facing the world has loosed upon that same long-suffering world a navalanche of articles minatory, pontifical or plain gloating: heaven knows how much energy has been used in the production and absorption of these writings. Most of the prophets of doom are concerned only with present fuels and present modes of extraction and use, and very properly — given their premises — come to deeply pessimistic conclusions. A broader approach leads to less apocalyptic conclusions: thus Surrey and Bromley, intheir contribution[1] to the University of Sussex's analyses of the Club of Rome's *The Limits toGrowth* , after concluding flatly that "there is simply no way of accurately assessing total world energy resources", rehearse the reasons for confidence that technological innovation can keep the crisis under control by developing new energy sources, new ways of using present ones more effectively, and ways of damping down the more profligate forms of energy waste.

Another voice of qualified optimism is that of Sir Alan Cottrell, the British Government's Chief Scientific Adviser. Last October he presented a survey lecture on materials and energy to members of the two principal American metallurgical societies, and this has now been published[2]. His theme is the central part that constructional materials must play in the more efficient extraction and exploitation of fuels and in the more economical use of the resultant power: in the process, he covers with his usual clarity and verbal economy a very wide range of devices for extracting energy from fuels, and in particular focuses attention on some aspects which are usually neglected, especially techniques of storing energy. His article, most of which is accessible to non-scientists, should certainly be read by industrialists and publicists concerned with the future of energy.

In passing, Sir Alan repeatedly refers to the suggestions which have been made for the widespread use of hydrogen as a vehicle for converting heat into an easily transort-able and storable form of energy. He quotes the electrochemist Bockris[3] as one recent proponent of this notion: it has also recently been analysed in *Scientific American* . Water would be electrolysed at nuclear power stations, presumably at night at periods of low electricity demand, and the hydrogen distributedthrough the natural gas pipe network, refrigerated and stored in liquid form in giant Dewar vessels. One way of using it, as Cottrell points out, is for direct conversion into electricity at high efficiency by means of Bacon fuel cells. A supplement to this proposal has recently been studied in the United States: liquid hydrogen proves to be anefficient and entirely pollution-free fuel for existing motor-car engines, and a serious analysis of the advantages and problems of using it in this way is now in progress.

As outlined, the proposal for a "hydrogen economy" involves hydrogen merely as a convenient intermediary, plus a means of absorbing electricity at periods of low demand and thus improving return on capital. But hydrogen could in principle be used to better effect. Cottrell points out the low overall efficiency, even now, of heat engines and their attached electric generators, and the benefits which will attend really effective large-scale direct conversion of heat into electricity or other forms of expressed or stored energy. The production of hydrogen by electrolysis is not efficient: apart from the inescapable thermodynamic limitations on the conversion of "nuclear heat" into electricity, large industrial electrolysers, according to Marchetti[4], are only about 50% efficient and very capital-intensive. Marchetti, who is in charge of materials research at the Euratom Laboratories at Ispra, Italy, makes this point in the course of a detailed account of the chemical hydrogen extraction process on which a team at Ispra has been working for some time. The Euratom method involves a cycle of several reactions. Several variants have been examined; the mark I cycle is as follows:

$$CaBr_2 + H_2O \xrightarrow{730^\circ C} Ca(OH)_2 + 2HBr$$

$$Hg + 2HBr \xrightarrow{250^\circ C} HgBr_2 + H_2$$

$$HgBr_2 + Ca(OH)_2 \xrightarrow{200^\circ C} CaBr_2 + HgO + H_2O$$

$$HgO \xrightarrow{600^\circ C} Hg + \tfrac{1}{2}O_2$$

The sum of these is $H_2O \rightarrow H_2 + \frac{1}{2}O_2$.

According to Marchetti, the thermodynamic efficiency of this cycle is about 85%, and the practical efficiency overall is at present about 55%

and still improving. Materials seem to pose no serious problems and recovery of the catalysts is straightforward. He goes on to a detailed analysis of the economics, in terms of capital costs and running costs assessed separately, of the generation and transport of hydrogen.

From information publicly available it would seem that at present all the running in this field is being made by American industry and by Euratom. It is to be hoped that the powers that be in Britain are at least maintaining a close watching brief at this stage.

The growing interest in the use of liquid hydrogen as a transportable, highly concentrated form of fuel can be regarded as a form of technology transfer. The American space programme has turned liquid hydrogen from an alarming speciality (used largely for nuclear bubble chamber research) into a familiar and not particularly dangerous commodity. But it is one thing for a large government agency to handle a material which was formerly thought unacceptably hazardous and quite another for the man in the street — or the roadside garage man — to do likewise. The precedents are not encouraging: liquid sodium is another material in the same category as liquid hydrogen. It has become familiar within specialized engineering circles because of its essential use as a coolant for certain types of nuclear reactor, and has been handled with complete safety for years past. Nevertheless, a brave attempt by a Union Carbide subsidiary to manufacture plastic-sheathed sodium cables for the transport of electricity has failed and the factory has had to be closed.

1. Surrey and Bromley, *Futures* **5**, 90 (1973).
2. Alan Cottrell, *Metall. Trans.* **4**, 405 (1973).
3. Bockris, *Science* **176**, 1323 (1972).
4. Marchetti, *Laser* (Liberal Party Science Magazine), No. 34, p. 11 (1973).

53

The Wind of Hydrogen and of Change Blew Gentle, Clean and Persistent at Miami...

Nature, **248**, 628 (1974).

Politicians and oil sheiks alike think of fuels as means of generating energy: another way of regarding them is as a means of storing energy. This approach was very much to the fore in the world's first conference on the "hydrogen economy", organised last month (March 1974) by the University of Miami, with the backing of the National Science Foundation, amid the lush pleasures of the Playboy Plaza at Miami Beach. The international meeting was attended by over 700 participants, more than four times the number expected even by the buoyant organisers. Those attending were divided between the aficionados — and one of the opening speakers truly presented the adherents of the hydrogen economy as members of a quasireligious cult — and a large number of mild sceptics who were sufficiently intrigued to want to see how seriously the topic should be taken. The most popular question while sipping drinks among the bunny girls was: "And why are you here?"

Hydrogen is not a natural fuel on earth. It must first be made, most conveniently from water, and then turned back to water, generating heat (or electricity directly in a fuel cell). Since neither generation nor combustion is perfectly efficient, the cycle $H_2O \rightarrow H_2 \rightarrow H_2O$ only makes sense from three aspects: it offers a compact means of storing and transporting energy which is not conveniently used at the site and time of liberation; it provides a possibility of creating a transport-fuel with very high energy content per unit mass; and it provides one possible substitute to pass through the insatiable natural gas pipelines of North America.

Nearly 100 papers (available in a preprint volume from the School of Continuing Studies at the University of Miami) were presented broadly on the following themes: generation of hydrogen by means of nuclear, solar or geothermal energy, and specifically by electrolysis or thermochemical cycles with catalysts, or by steam reforming of carboniferous fuels; storage of hydrogen as liquid or, more especially, in thermally decomposable hydrides; transport of hydrogen through pipelines, including problems of embrittle ment by hyrogen; hydrogen as a fuel for cars and aircraft; the relative cost of hydrogen and other fuels, particularly methane, for various uses. Indeed, the conference was notable for the scrupulous care devoted to the economics of the various patterns of hydrogen production and use. The many attempts at costing covered a moderately wide spectrum, and one author's assumption was on occasion disproved by another author's technical innovation. Thus, one of the numerous papers on thermochemical cycles, by R. E. Chao (University of Puerto Rico) and K. E. Cox (University of New Mexico), concluded that "thermochemical hydrogen via a series of closed chemical cycles and nuclear heat would cost $1.50-2.35 per million BTU, whereas hydrogen from electrolysis/nuclear power would cost $4.60-6.28 per million BTU" (compared with present jet fuel costs of $1.70 per million BTU). But Chao and Cox assumed electrolyser capital costs at $200 per kW, whereas L. J. Nuttall, A. P. Fickett and W. A. Titterington of the General Electric Company (GE), in their very impressive paper on hydrogen generation by electrolysis with solid polymer electrolyte, estimated capital costs which might be as low as $60 per kW and overall hydrogen costs in the range of $2-3 per million BTU (excluding liquefaction costs), which is distinctly interesting in the medium term. The GE paper was a good example of high technology transfer, since it described a thoroughly engineered process originally designed to make tonnage oxygen for the space programme, with hydrogen going to waste; this could easily be turned on its head. The GE technique effectively separates the two gases, permits extremely high current densities and, very important, gets by with exceptionally low loadings of noble metal catalyst. Ordinary commercial electrolysers cannot afford to use catalyst and thus have to operate very inefficiently.

The numerous papers on thermochemical cycles left an overall impression of a technologically very difficult method, dependent on extremely efficient recovery of catalyst and with many uncertainties as to overall efficiency of energy conversion. It is also remarkably difficult to decide an on optimum series of chemical cycles for investigation: even a computer-aided search of possible cycles, by J. L. Russell and J. T. Porter of the General Atomic Company, gave conclusions which are hard to evaluate. One listener was left with the impression that electrolysis scores on points over thermochemical cycles, until the end of the century at least. This view was not weakened by a very interesting paper by J. B. O'Sullivan and others of the United States Army, who described a detailed and hitherto unpublicised feasibility study of a mobile hydrogen generator for field use, centring on a thermochemical cycle. The full details of this are to appear in the final proceedings of the conference, to be published by Plenum Press.

Opinions were violently divided on hydrogen as a fuel for internal combustion engines. E. M. Dickson and colleagues of the Stanford Research Institute presented a closely argued critique, with pessimistic conclusions, of the use of hydrogen to fuel 'personal vehicles'. An enthusiast, R. E. Billings of the Billings Energy Research Corporation (many small companies were represented) however described in optimistic detail his own experience of operating a fleet of hydrogen-fuelled motor-caravans, and he also claimed considerable operating data for a converted passenger car working interchangeably with hydride or cryogenic storage.

Papers dealing specifically with hydrogen storage in cars were particularly interesting. Several discussed the detailed engineering design of hydride storage beds, mainly with reference to $FeTiH_{1.6}$ and MgH_2. The first of these is heavy but decomposes at low temperatures; the second is light but needs to be heated to over 600°C. Billings explained how an optimum (i.e., lightest feasible) combination of the two hydrides could be used, with as much MgH_2 as could be sufficiently heated by the exhaust gases of the engine, while hydrogen was liberated from the other hydride by means of the engine-cooling fluid. Considerable and sophisticated attention was paid to the engineering design of hydride beds for fastest possible charging and discharging, for instance by G. J. Powers and D. C. Cummings of MIT. The kinetics of hydride formation or decomposition may be limited either by heat flow or by hydrogen diffusion; the consensus appears to be that in the best designs, heat flow is limiting. Complete charging of a hydride bed in five minutes was claimed by one speaker; this is unlikely to be bettered.

Overall, the technology of hydrogen-fuelled motor cars seemed fairly promising; as several speakers pointed out, however, hydrogen-fuelled cars (like hydrogen as substitute natural gas) would probably require a huge discontinuous change in distribution over the whole of a large area all at once; it could scarcely come by easy stages.

Methanol, which can be made by a hydrogen route, received attention as a possible automotive fuel. In an impressive paper modified by some last-minute findings, T. B. Reed and R. M. Lemer of MIT dealt with the potential for progressive lacing of petrol with synthetic methanol, as methanol gradually becomes cheaper relative to oil. The prospects are promising, according to this work, that increased methanol content will so improve anti-knock properties that the lead content can be considerably lowered or perhaps even eliminated. The conflict of all-

or-nothing versus progressive introduction of new fuels was one aspect of the "social" approach to the hydrogen economy which several speakers essayed. Dickson's paper from Stanford was outstanding in this connection. He analysed the magnitude of the impact of various changes inseparable from a hydrogen-fuelled car economy and also examined which groups of people (drivers, mechanics, fuel dispensers, fuel distributors, fuel manufacturers) were affected. His historical parallels — for instance, the disincentive to the spread of diesel-fuelled passenger cars which he identified with the mild inconvenience of locating suppliers of diesel fuel — lead him to argue strongly that while quite major changes in engine design, such as the Wankel engine, can be quite readily accommodated, relatively modest changes in nature of the fuel, and especially of fuel distribution, will constitute major blockages.

One advantage of the kind of analysis offered by Dickson is that it can alert technologists to needs that otherwise might be underestimated or even ignored. If it is accepted that the introduction of a hydrogen-only car would come up against particularly intractable socio-political resistance, then there is a strong incentive to develop an engine that can run on either of two alternative fuels — such as the hydrogen/methanol engine described at the conference by R. R. Adt of Greenwell and M. P. Swain of the University of Miami. Technology and economics are but two sides of a triangle, of which politics forms the base.

The long paper by T. C. Cody of West Virginia University on the hydrogen economy and (American) law was particularly thought provoking. Cody foreshadows increasing "federal preemption" over the state in safety and environmental legislation and a progressive sharpening of legal limitations on energy use: eventually he foresees a law of allocation to replace the present "crazy quilt of contradictions, cross-purposes, omissions, overlap and duplication".

On hydrogen for aviation, opinions were also divided. R. A. Lessard, an economist of the United Aircraft Research laboratories, in a separately circulated paper, was sceptical on purely economic grounds, whereas P. F. Korycinski and D. B. Snow of NASA were optimistic on the basis of a detailed technical analysis. The larger the aircraft and the greater the range, the greater is the economic advantage of using liquid hydrogen. One listener at least was left with the clear impression that aviation will be the first major user of hydrogen, well before the century is out. This view was shared by a severe critic of the hydrogen economy, P. N. Ross, a vice-president of Westinghouse, who was utterly convinced that nuclear electricity, distributed as such, would indefinitely defeat the use of hydrogen — except for aircraft.

There is no space to do justice to the group of papers on solar and geothermal energy (the latter seems particularly promising in an

American context). The conference showed very clearly its recognition that hydrogen is merely a means of storing energy, not creating it. Some new, cheap solar cells were developed and evaluated, the engineering aspects of a mile-square servo-controlled mirror array and boiler tower were analysed, there was much to-do about large floating platforms making use of thermal gradients in the oceans (a form of indirect solar energy) and wind power came in for some detailed analysis, not least in connection with a self-contained domestic system using only established technology, including energy storage by electrolysis and later use of the hydrogen. This intriguing study was by L. W. Zelby of the University of Oklahoma, where the wind is gentle but persistent.

Indeed, the wind of hydrogen and of change blew gentle, clean and persistent at Miami. The fact that hydrogen is almost pollution-free was much commented on; the name of the Environmental Protection Agency was much mooted and an International Association for Clean Energy was founded at the conference. There was the sniff in the air of American technology on the move; much more will be heard of hydrogen before long.

Fuel prices and the energy crisis

Fuel	*Price before the energy crisis ($ per million BTU)*	*Estimated 1974 price*
Natural gas	*0.25*	*0.70*
Fuel Oil	*0.32*	*1.80*
Coal	*0.20*	*0.58*

54

Between the Devil and the Deep Blue Sea

Nature, **266**, 106 (1977).

In the anguished debates on the future if any, of nuclear fission power, various recognisable strands emerge among the opposition. Some pin their faith on the improved technological exploitation of familiar fossil sources, principally coal; others would drive us headlong into the solar economy, often conceived in terms of "small is beautiful" and a rural life of log-cabin simplicity; yet others accept the need for high technology on a very large scale if the sun is to yield domestic warmth, electricity and liquid fuel on an adequate scale.

All three groups, however, have one characteristic in common. When faced by hard statistical or technological criticisms, they declare that we need only keep going for a few decades and then the boundless energy "from sea- water" — that is, nuclear fusion — will become a reality and all our worries will be over. Although they do clearly recognise that the method is not altogether simple, even the most resolute proclaimers of radioactive doom seem to have convinced themselves that fusion power will be gentle and on a homely scale, and that radioactive pollution will be a thing of the past when dirty fission gives way to clean fusion. Probably the fact that fusion takes place in the sun and that the sun is the emblem of cleanliness and of life itself, has something to do with this touching faith. We are all to some degree subject to the ad-man's technique of fallacious association.

There is no dearth of popularisations of this very difficult and complex topic, but they all concentrate on two aspects of fusion — how to confine the plasma, and how to heat it to ignition temperature. Even within these limits, they tend to concentrate on scientific principles and not on engineering realities, whereas the current criticism of nuclear fission concentrates on real and alleged engineering weaknesses. Nuclear fusion has progressed far enough now towards the design of a self-sustaining reactor to make it both possible and necessary to see the problems in realistic terms. The plasma confiners — who indeed have a ferociously difficult task — until recently have hogged all the attention, to the neglect of the problems of energy balance (achieving a positive net energy output), container strength and fatigue damage, lithium supply, reactor cooling, injection of gas into the plasma without quenching it, and especially the radiation damage inside the reactor itself.

That this imbalance is now being repaired is primarily due to the efforts of the Energy Research and Development Administration (ERDA), which has complete control of fusion research in the United States. ERDA has prepared a detailed R&D plan extending to 1990, and in outline to the end of the century. ERDA's achievement is to have set out this programme without committing itself to any particular design of reactor, and yet to have identified the central problems which will have to be faced irrespective of the design. To do this, ERDA's predecessor in 1967 invented the concept of the "reference design". This is an outline based on scientific principles, as presently understood, set out in such a way as to identify, as quantitatively as possible, the whole range of engineering problems that will need to be mastered. Such a reference design is particularly illuminating when applied to the doughnut-shaped Tokamak, which in purely scientific terms is today probably the most hopeful contender among the rival approaches.

On the basis of such a reference design, the problem of securing the deuterium (by electrolysis of water?) and tritium from a metre-thick breeding blanket of molten lithium surrounding the first wall of a doughnut-shaped plasma device, can be analysed, not least in terms of lithium availability; lithium is a scarce element. One day it will also be necessary to include in the energy balance equation not only the energy cost of extracting tritium from irradiated lithium but also the electricity costs of electrolysing water to extract deuterium, a calculation which never seems to be reported.

Probably the most alarming engineering problem is that of radiation damage. In the much-criticised fission reactor, neutrons, gamma rays and fission particles remain inside the reactor volume, and indeed the fission products remain inside the nuclear fuel itself. In a Tokamak, all the neutrons, moving with a kinetic energy of some 14 MeV, beat upon the first wall: the scale of this bombardment dwarfs anything in a fission reactor. The wall can be sputtered away, especially by the novel hazard of 'chunky sputtering' which involves the physical removal of thousands of atoms at a time. [That last phenomenon proved to be a false alarm, soon after this article was published].

A great deal of research is now in progress on this group of problems in the United States (some also in Britain), and considerable advances have been made in improving the radiation resistance of high-strength refractory alloys. But it is certainly an error to regard fusion reactors as generating little radioactivity: there will be a great deal of induced radioactivity, though not from fission products, and it will be just as crucial as in the case of a fission reactor to contain and seal off this radioactivity from the outer world. (It seems likely that this hazard will not be long-lived). Apart from this, there will be just the same hazards of pinholes developing in inaccessible locations as have been found with prototype fission-breeder reactors, leading to deterioration in cooling efficiency or vacuum quality.

Fusion may prove to be a practical source of energy — the issue is now as much in the hands of the engineers and metallurgists as of the plasma physicists. If it comes, it will be an immeasurable boon but it will also bring with it problems of pollution, energy imbalance during construction, resource depletion, sabotage and potential stoppage such as we find with every other source of energy, including solar techniques such as the large-scale fermentation of timber to make liquid fuel.

For Blake, energy was a mark of the divine. Milton knew better: Satan has energy too. Whether the eternal flames are fed by coal, oil, uranium or deuterium is a question which must tax the physical theologians of the future.

55

Solar Energy Storage and Conversion by Hydrogen Cycle

Nature , **276**, 665 (1978)

All forms of 'current' energy — sun, winds, waves, tides — are intermittent and variable and their use for large-scale and domestic purposes alike demands provision of energy storage. This may well turn out to be more difficult than and just as capital-intensive as the devices for absorbing and converting the energy in the first place. The recent controversy in *Nature* concerning the storage requirements associated with a system of windmills indicates the crucial importance of the storage capacity if steady power is to be generated from variable input.

Most domestic solar heating installations to date are intended largely or wholly for water-heating, and the thermal mass of an insulated hot-water tank provides all needful storage capacity. If, however, space-heating is proposed as well, or even electricity generation, then a more sophisticated storage facility is needed. Heat can stored in pebble beds, water tanks or salt baths and retrieved by forced heat exchange when required. However, space is expensive and a high storage capacity per unit volume (specific capacity) is therefore at a premium. A high specific capacity implies exploitation of chemical reaction energies as distinct from simple specific heats, and it is in this connection that metal hydrides have received a good deal of attention in the past four years.

A number of metal hydrides (TiFe, $LaNi_5$, Mg_2Ni and Mg_2Cu are the most studied) have convenient dissociation pressures and temperatures, substantial heats of dissociation and they can store more hydrogen per unit volume than can a liquid-hydrogen Dewar vessel[1]. A number of laboratories, of which Brookhaven National Laboratory has been the most active, have been examining the metallurgical and engineering features of hydride storage installations, partly with automotive uses in mind. A recent paper from Argonne National Laboratory[2] presents a conceptual study of a system using such hydrides for storage of heat and its use in adjustable proportion for domestic heating, air-conditioning and electricity generation. It is a welcome publication, since much work in this field is issued in inaccessible reports or expensive conference proceedings — what might loosely be termed scientific *samizdat.* All 12 references at the end of this paper are to unpublished reports, house journals or conference proceedings, and this practice leaves little scope for the serendipitous browsing that so often generates new scientific ideas. For this reason, the existence of regular journals like the *International Journal of Hydrogen Energy* is particularly welcome; in a field like this, the absence of a highly specialised journal is apt to be associated with entrenched scientific *samizdat* .

The Argonne apparatus, in the simplest of its several forms, involves a solar collector with concentrator that can warm a heat-transfer fluid ot 140°C. The warm fluid then heats up a primary $LaNi_5H_6$ bed, releasing hydrogen at about 50 atm. The released hydrogen is cooled to 50°C and expanded to a pressure of 50 atm or less. The cool, lower-pressure hydrogen is then converted to hydride in a secondary bed (a cheaper bed using mischmetall [Mm], a mixture of rare earth metals instead of lanthanum, is proposed). The heat released during expansion and hydration is absorbed in a separate loop of heat-transfer fluid, and this heat (at 50°C) can be used for space-heating. Hydrogen is released from the cooled $MmNi_5H_6$ secondary bed by lowering the pressure to about 8 atm and can be recombined with the primary $LaNi_5H_6$ bed at 50°C at the same pressure. That bed is then reheated by solar energy and the cycle recommences.

This represents a simple non-storage cycle. In fact, the storage bed would be kept hydrogenated till nightfall and the apparatus would then be operated in heat-pump mode; the primary bed is kept at 50°C by manipulating heat-transfer fluid and is used as a source of space heat, while the secondary bed can progressively dehydrogenate at low pressure and 0°C; this low temperature is achieved by absorbing heat from the environment.

To achieve continuity in operation (as would be required for continuous air-cooling or for electricity gener ation) at least two primary and two secondary beds would be used, with appropriate switching of hydrogen and heat-transfer fluid flows so that each undergoes a staggered cycle: absorption, heating, desorption, cooling. The cycle length is presumably limited by maximum heat-transfer rates at one extreme and by low demand for domestic heat at the other extreme. A variant of this cycle allows the apparatus to be used for air-conditioning instead of space-heating.

A more complex system involving three beds of each kind would permit a proportion of the absorbed solar energy to be used for electricity generation. (Warm hydrogen would be passed through a turbo generator: the alternative of oxidising the hydrogen to water in a fuel-cell would be inappropriate in a closed cycle system). The authors undertake a detailed assessment of realistic efficiencies to be expected, allowing for heat losses involved in heating and cooling the storage beds: the maximum theoretical efficiency is 28% (electric power as fraction of solar heat absorbed by the primary bed), the expected realistic efficiency is 16.5%. If the apparatus is used simply as an air-conditioning (cooling) device, the heat-pump mode ensures that, at ambient temperature, for each mole of hydrogen about 3.3 times the dissociation energy can be extracted as useful cooling power.

Nothing is said about economics. The high cost of mischmetall and even higher cost of pure lanthanum would seem to place a question-mark against the particular hydrides chosen, in spite of their particularly convenient dissociation temperatures.

The Argonne design prompts a number of reflections. The apparatus would be of very considerable complexity. Each bed would be associated with several heat-transfer loops, pumps and pressure valves, and there would need to be an elaborate automatic control installation to switch both hydrogen and heat-transfer fluid into appropriate circuits at appropriate times, to secure the staggered cycles and also in response to changes of incoming solar power level, space-heating or cooling demand, hydride temperature, hydrogen pressure and fluid temperature in various parts of the apparatus. This is quite apart from the control provisions needed to keep electric output constant in the face of varying hydrogen temperatures and pressures. There is a widespread illusion that renewable energy sources such as sun and wind are necessarily associated with primitive or intermediate levels of technology: people are apt to think in terms of solar cookers for use in Indian villages or of water-heaters for swimming-pools or for Israeli dwellings. In fact, if solar power is ever to be more than a negligible fringe energy source, plainly it must be accompanied by very advanced technology. This is true both of the initial capital input (such as the manufacture of silicon solar cells) and of the maintenance of complex installations, which will place high demands on initial design (perhaps in the form of replaceable modules), on fault diagnosis and on high reliability in manufacture. Such things will certainly be "mills" in William Blake's sense, though bright and beneficial rather than dark and satanic.

In the very early days of domestic electrical power, a century ago, there was a brief sharp argument as to whether separate electric generators were to be installed in the basement of each home, or central generation and distribution was preferable. One may wonder whether, before another century is out, this argument will be re-opened in sunny parts of the world. Meanwhile, the most promising application, in the short term, of installations like the Argonne design would be for air-conditioning in hot, isolated and fuel-deficient places like North western Australia. Such an air-conditioning facility, built into dwellings and communal spaces, may well be the factor that determines whether or not labour can be attracted to work in mines in such inhospitable places.

1. Hoffman *et al.*, *Int. J. Hydrogen Energy* **1**, 133 (1976).
2. Green, Schreiner and Shaft, *Int. J. Hydrogen Energy* **3**, 303 (1978).

56

Towards Photoelectrolytic Power

Nature, **302**, 294 (1983)

Photoelectrolysis, once an electrochemist's plaything, is close to competing with the photovoltaic silicon cell as a means of converting sunlight into power. This development is due to two kinds of advance: an improvement in the requisite semiconducting photoelectrode, and subtle treatment of the electrode surface in order to stabilize it against deterioration. A short while ago, an article in *News and Views* [1] described some of the problems associated with electrodes, especially if they are polycrystalline. One way to solve the problems, described in the same issue[2], is to use a non-aqueous electrolyte; other possibilities compatible with an aqueous electrolyte, and indeed with water electrolysis, are outlined here.

Many of the key inventions in photoelectrolysis have come from the Bell Laboratories, and so repeat the history of the transistor and the photovoltaic cell. Brattain and Garrett of Bell Labs, in 1955, laid down the now classical tbeory of the illuminated semiconductor-electrolyte junction. The idea of using an illuminated semiconducting electrode as either anode or cathode in an electrolytic cell to produce electric power or chemical products arose naturally from their work, but the very rapid fall in efficiency of electrodes under irradiation was not solved until 1972, when Honda and Fujishima reported, in this journal[3], the stability of (n -type) TiO_2 photoanodes. Unfortunately, such electrodes are useless for converting sunlight as the 3-eV energy gap is much too large. Efficient solar conversion requires gaps in the range 1.3 ± 0.3 eV to absorb light in the λ range 1 000 ± 250 nm.

In 1975, Adam Heller joined Bell Labs to initiate a research programme on photoelectrolysis. At that time it was believed that electrodes capable of giving efficient conversion of sunlight could not do so for long, unless astute chemical counter measures could be applied at the electrode surface. Success has been achieved at Bell by combining several strategies: photocathodes such as p –InP (1.3 eV), p –GaAs

(1.4eV) or *p* –CdSe (1.5 eV) are intrinsically stabler than (*n*–type) photoanodes[4]; surface recombination of electrons and holes can be reduced by chemical modification of the electrode surface (anodic formation of a thin hydrated indium oxide layer on InP, or absorption of Ru^{3+} ions at InP or GaAs surfaces); if electrolysis of water is the aim, the incorporation of thin translucent layers of noble metal helps. In these ways[5-7], 'solar' efficiencies of up to 16 per cent have been achieved, well up to photovoltaic levels. Such electrodes can be used either in spontaneous cells (without externally supplied electric power) to make electricity, or in photoassist-ed externally biased cells to electrolyse water; efficiency can be defined either in terms of electrical power saved or of the power available in an ideal fuel cell from the hydrogen evolved, and the resulting efficiencies differ[8]. One can even combine two photoelectrodes, *p* and *n* type, in a single cell, for example to electrolyse HBr into H_2 and Br_2 [9].

Honda and Fujishima's cell used single-crystal TiO_2, and the early Bell Labs cells were also based on single-crystal semiconductors. One of the principal advances at Bell was to establish that the same absorption strategy used to improve electrode kinetics — to slow down electron-hole recombination — at a monocrystal surface will also lead to stabilization of grain boundaries against recombination: thus, Ru^{3+} ions improve the efficiency of a polycrystalline *n* –-GaAs photoanode by a factor of 4, to the point where such electrodes can be used for practical photoelectrolysis[10]. This represents a fundamental advantage over photovoltaics, where it has hitherto proved impractical to slow dopant diffusion along grain boundaries in polycrystalline silicon to the point where good *p–n* junctions can be made and the lesser cost of polycrystalline material effectively exploited (see refs 4, 8).

Until recently[8], it has not been possible to electrolyse water by either spontaneous or photoassisted photoelectrolysis when a photoelectrode with energy gap ≈ 1.3 eV has been used. The inefficient *n* –TiO_2 photoanode can, however, be so used[11] and is, moreover, highly stable chemically (in spite of being an anode and thus more liable to photocorrosion than a cathode). There is therefore much interest in any procedure which holds promise of increasing the photoelectrochemical efficiency of *n* –TiO_2.

This, it appears, can be done in one of two ways: either by departing increasingly from stoichiometry in the direction of oxygen deficiency — TiO_{2-x} [12]; or by 'alloying' TiO_2 with VO_2 [13,14]. Experiments have been done on both single crystals and sintered polycrystals, and in these stable materials, grain boundaries cause little deterioration relative to monocrystals (though it is not clear why electron-hole recombination at grain boundaries is insignificant, even in the alloyed materials with trolyte and 335-nm wavelength, increased in efficiency up to 21 per cent for $x = 7 \times 10^{-4}$, then decreased. The decrease was attributed to enhanced surface recombination. It should be pointed out that the recorded 21 per cent efficiency is for monochromatic light, not for a solar spectrum.

The $Ti_x V_{1-x} O_2$ alloys allow the energy gap to be reduced, down to 1.95-2.04 eV (measurements differ), without, it appears, any loss of chemical stability[13,14] and without deterioration of efficiency when a polycrystal replaces a single crystal. ($Ti_x Nb_{1-x} O_2$ electrodes were, however, no better than pure TiO_2.) A gap of 2 eV is close to the range for optimum solar efficiency and could be used for a reasonably efficient solar cell. The study of alloyed semiconducting oxide electrodes is only in its infancy, and it may well prove possible to lower the energy gap further without serious loss of chemical stability.

The technological aspects of large-scale photoassisted electrolysis of water have not been examined but it can be most efficiently performed at high temperature, and thus under enhanced pressure, with appropriate electrocatalysts to reduce the hydrogen overvoltage, in devices where multiple layers of anodes and cathodes are interleaved. It is not obvious how such an apparatus could be adapted to permit access of sunlight to the anodes, and if electrolysis is done at ambient temperature and pressure, the loss of efficiency may conceivably outweigh the benefit derived from photoassistance. A study of the combined efficiency of water electrolysis with photo-assistance under different conditions would seem to be the next stage in the approach to the capture of sunlight with the aid of semiconductors.

1. A. Hamnett and S. Dennison, *Nature* **300**, 687 (1982).
2. C.M. Gronet and N.S. Lewis, *Nature* **300**, 733 (1982).
3. K. Honda and A. Fujishima, *Nature* **238**, 37 (1972).
4. A. Heller, *Acc. Chem. Res.* **14**, 154 (1981).
5. A. Heller, B. Miller and F.A. Thiel, *Appl. Phys. Lett.* **38**, 282 (1981).
6. H.J. Lewerenz, D.E. Aspnes, B. Miller, D.L. Malm and A. Heller *J. Am. Chem. Soc.* **104**, 3325 (1982).
7. E. Aharon-Shalom and A. Heller, *J. Electrochem Soc.* **130** (at that time in the press).
8. A. Heller, *Solar Energy* **29**, 153 (1982).
9. C. Levy-Clement, A. Heller, W.A. Bonner and B.A. Parkinson, *J. Electrochem. Soc.* **127**, 90 (1980).

10. W.D. Johnston, H.J. Leamy Jr. and B.A. Parkinson *J. Electrochem. Soc.* **127**, 90 (1980).

11. M.E. Gerstner, *J. Electrochem. Soc.* **126**, 944 (1979).

12. J. Gantron, P. Lemesson and J.F. Marucco, *Faraday Disc.* **70**, 81 (1980).

13. P. Lemasson, M. Etman, J. Gantron and J.F. Marucco, *Proc. 4th Int. Conf. on Photochemical Conversion and Storage of Solar Energy*, Jerusalem, August 1982 (in the press).

14. T.E. Phillips, K. Moorjani J.C. Murphy and T.O. Poehler, *J. Electrochem. Soc.* **129**, 1210 (1982).

57

Smooth Tritium for Laser Fusion

Nature, **335**, 399 (1988)

The Matthew principle is a well founded informal generalization in science and in life: the large will grow larger, the small will vanish. It is named after the disconcerting biblical passage[1], in which St Matthew recounts the parable of the talents: "To every person that has something, even more will be given and he will have more than enough; but the person who has nothing, even the little that he has will be taken away from him". In materials science, the standard exemplar of the Matthew principle is the process termed *Ostwald ripening*. If an alloy, say, contains a distribution of spherical precipitates of diverse sizes, then, so long as the constituents of the precipitates are soluble in the matrix and can diffuse, the larger spheres will grow and the smaller will shrink, following a well established $t^{1/3}$ dependence on time. In a new study[2] of a non-metallic system, Hoffer and Foreman demonstrate that there are circumstances under which, contrary to the Matthew principle, initial variations are evened out. They have used radioactive tritium in the study, which has important implications for nuclear physicists exploring the use of frozen tritium pellets for nuclear- fusion power generation.

Ostwald ripening occurs because the solubility of the precipitates is enhanced near a highly curved interface, that is, near a small sphere. The kinetics depend on solubility, diffusivity, the specific energy of the interface and the temperature[3]. The universality of the Ostwald ripening phenomenon has been dented by the observations of ultrafine particles in certain alloys which do not ripen at all, for reasons which are still disputed, and also by the recent analysis of the role of elastic stress at coherent interfaces, which can result in a preference for a particular precipitate size over smallerand larger ones[4]. But in general, theMatthew principle leads to the enhancement of initial variations and irregularities.

The new study by Hoffer and Foreman[2] refers to the radioactively induced sublimation in solid tritium. Tritium is radioactive, generating low-energy electrons, which are absorbed within 20 μm in solid tritium, depositing heat and ^3He. A slab of solid tritium, the outer surfaces of which are kept at constant temperature, would therefore have a parabolic temperature distribution, with a maximum at the plane in the centre. If a thin slice in such a slab is converted into vapour, so that the original slab now consists of two unequal, thinner slabs (see figure), heat should betransferred across the vapour-filled gap from the thicker slice (where more heat is generated) to the thinner by sublimation. Assuming that there is no obstacle to the flow of vapour, then because the rate of heat generation, $\dot{q}$, in the solid is everywhere uniform, the rate of decrease of the 'excess thickness', δ, should be given by $\delta = \delta_0 \exp(-t/t_{min})$, where δ_0 is the initial value of δ and $t_{min} = H_s/\dot{q}$; H_s is the heat of sublimation. The end result should be a pair of slabs of equal thickness.

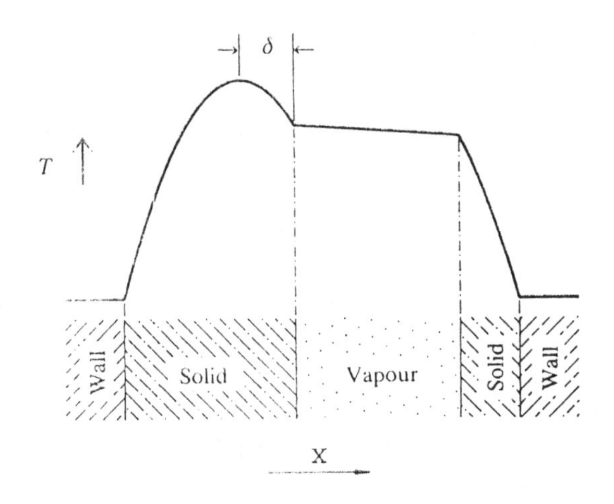

Temperature profile, described in the text, for two solid, self-heating layers of tritium. (From ref. 2)

Hoffer and Foreman tested this prediction experimentally, using a cylindrical geometry. They froze freshly prepared liquid molecular tritium in a horizontal cylindrical vessel (5.74 millimetres deep and across) at constant temperature with end windows: the solid tritium initially has the original meniscus form of the liquidand thus has the shape of a concave lens on the lower side only of the cylinder. Progressively, sublimation converts this into auniform annular coating of the entire cylindrical wall, the kinetics following the predicted relation, with t_{min} only a few percent greater than the predicted value of 14.4 minutes at 19.6 K. When the same experiment is done with 7-day-old tritium, t_{min} is more than six times greater; this Hoffer and Foreman attribute to the interference with tritium vapour flow caused by the stagnant pool of helium, the radioactive decay product. Sublimation of the self-heated radioactive solid thus leads to removal of an initial geometrical irregularity.

This process is somewhat reminiscent of the removal of protuberances and grooves from a mechanically polished metal surface by electrolytic polishing. For this, the rate of migration of ions through a thin, viscous electrolyte layer at the metal surface is greater where a protuberance causes local reduction of the thickness of the viscous layer. An analogous process is polishing by simple differential evaporation in a vacuum, where a protuberance evaporates faster than a groove[5].

These findings are relevant to the design of fuel targets for nuclear-fusion power based on inertial confinement (see the recent *News and Views* article by R.S. Pease[6]). The demands on the dimensional uniformity and symmetry of such targets are extreme and the difficulties of making them are correspondingly great[7]. A.J. Martin, D. Musinski and R.D. Simms proposed in 1985, in unpublished work cited by Hoffer and Foreman, that the sublimation effect, which they predicted, could be used to make a deuterium-tritium target in the form of a highly uniform, spherical microannulus. The possibilities are discussed in a forthcoming paper by Hoffer and Foreman[8]; they look promising and this approach would presumably be much less troublesome than others: for example, absorbing a uniform annulus of liquid deuterium-tritium in a porous polymer foam within a micro-balloon. A fusion target consisting of a frozen, uniform microannulus of deuterium-tritium could also easily withstand the mechanical effects of its rapid insertion into the focus of the initiating lasers.

1. Matthew, **25**: 29.
2. J.K. Matthew and L.R. Foreman, *Phys. Rev. Lett.* **60**, 1310 (1988).
3. J.W. Martin and R.D. Doherty, *Stability of Microstructure in Metallic Systems,* 173-221 (Cambridge University Press, 1976).
4. P. Voorhees and W.C. Johnson, *Phys. Rev. Lett.* and *Acta metall.* in the press (at the time).
5. R.S. Irani and R.W. Cahn, *Metallography* **4**, 91 (1971).
6. R.S. Pease, *Nature* **335**, 206 (1988).
7. R.W. Cahn, *Nature* **311**, 408 (1984).
8. L.R. Foreman and J.K. Hoffer, *Nuclear Fusion* , in the press (at the time).

Materials for Nuclear Power

58

Precipitate Instability

Nature , **241**, 16 (1973)

It is not at all surprising that an unprecedented research effort is being devoted to a form of radiation damage that is specific to fast reactors for, even though nuclear power, under pressure from environmentalists, is unfashionable, those persons responsible for national energy policy know perfectly well that in the long run most power must come from nuclear reactors, and the fast reactor in particular.

Radiation damage in fast reactors is a consequence of the fast neutrons and the high fuel temperatures : fuel swelling is caused by the formation of voids, which are themselves formed by the selective agglomeration of excess lattice vacancies resulting from neutron damage. When a neutron starts a collision sequence which causes a number of atoms to be knocked out of the crystal lattice, equal numbers of vacancies and interstitials result. If, as is all too apt to happen, the interstitial atoms migrate faster than the vacancies, then the former quickly disappear at 'sinks' and the vacancies are left high and dry; they are then likely to combine to form microscopic cavities or voids, and the fuel or the material of the fuel envelope swells dangerously. This has been a serious problem, but because of extensive research since 1967 it is now well on the way to solution.

One of the principal remedies is to use alloys containing a fine dispersion of precipitates; migrating interstitials are in some way arrested when they impinge on a precipitate particle, and are held captive there until they are neutralized by a wandering vacancy. So, in the end, almost all the interstitials recombine with vacancies and few vacancies are left to form voids. Thus, some nickel-base alloys of the Nimonic type have been found particularly suitable for the manufacture of fuel cans, for the fine precipitates of γ' (ordered Ni_3Al) impede swelling in the manner described. For this purpose, it is necessary for the precipitates to be particularly finely distributed (to increase the probability of interstitial capture), and to stay that way in service. It is this last requirement which is difficult, because all precipitate populations are subject to the law of Ostwald ripening — that is, operation of the Matthew Principle, or the growth of the large at the expense of the small.

A group of investigators at Harwell have now systematically examined, both theoretically and experimentally, the behaviour of precipitate populations in a radiation field. The Matthew Principle is counteracted by the steady destruction of precipitates by neutron damage, and the diffusion of solute atoms into the matrix at a rate which is itself accelerated by the presence of excess vacancies. From their theoretical calculations, Nelson, Hudson and Mazey[1] find that big particles redissolve faster than small ones, and that this is true whichever of two alternative mechanisms of resolution operates.

The final conclusion is that in the presence of steady neutron irradiation, large particles dissolve preferentially but small particles on balance still grow, so that a dynamic equilibrium is established and most particles finish up close to a preferred size, which is quite small. Irradiation thus effectively neutralizes Ostwald ripening. This conclusion is confirmed by experimental observations on a Nimonic-type alloy containing coherent γ' particles, but not by observations on another kind of nickel alloy containing incoherent precipitates. The last of these are subject to normal coarsening, because (it seems) one of the mechanisms for irradiation-induced resolution of precipitates, which requires mass transport across the particle/matrix interface, does not work satisfactorily across incoherent (high energy) interfaces.

1. Nelson, Hudson and Mazey, *J. Nucl. Mater.* **44**, 318 (1972).

59

Beaded Bubbles

Nature, **267**, 671 (1977)

An infinite monocrystalline element containing a second element in solid solution, not liable to any phase transformation, tends to a spatially uniform distribution of the dissolved element, and diffusion will persist until that stable terminal state is attained. That cautious opening sentence incorporates several provisos: if the solid consists of distinct crystal grains or, being less than infinite, is bounded by free surfaces, complications ensue. In general, solutes tend to segregate either to or away from free surfaces or grain boundaries; improved methods of microanalysis have allowed a considerable body of experimental information to be assembled concerning this process of 'Gibbsian' segregation[1,2]. Hondros's extensive experimental work has led him to the conclusion[1] that solutes with a small equilibrium bulk solubility (usually associated with very different solvent and solute atomic sizes) are particularly apt to segregate to interfaces and surfaces, thereby relieving lattice strain in the solvent. As Westbrook[1] points out, surface segregation is not restricted to external surfaces; it can apply equally to internal pores, such as are found in imperfectly sintered powder compacts. Westbrook emphasises the role of the binding energy that holds vacancies and solute atoms together; if this is high, any excess vacancies diffusing to a pore must drag solute atoms along with them.

Materials scientists responsible for the design of nuclear fuel elements are currently much concerned with pores — *voids* is the preferred term — in neutron-irradiated metals and ceramics. During irradiation (especially of the metal used to encapsulate the fuel) interstitial atoms and vacancies are formed in equal numbers and diffuse away to 'sinks', predominantly dislocations and grain boundaries. These sinks attract interstitial atoms more effectively than vacancies and the surplus vacancies then assemble to form small voids, visible in the electron micrscope. The voids cause swelling of the material, a disease which must be kept within strict bounds for safety reasons. The principal strategy for minimising swelling is to choose suitable base metals and to maximise the concentration of effective sinks for interstitials and vacancies alike. Now, another strategy has been identified.

Some years ago, Okamoto and Wiedersich at the Argonne National Laboratory[3] observed indications that solute elements segregate to radiation-induced voids in a stainless steel. The evidence was indirect — micrgraphic features indicating pronounced lattice distortion alleged to accompany the segregation — and it was not possible to identify the segregating species. The investigators suggested that it was interstitial solute atoms generated by the irradiation which segregate and lead to the high lattice distortion observed in the electron microscope. Thus Hondros's view that segregation acts to reduce lattice strains is not always in accord with observation.

The whole matter has now been clarified by an elegant set of experiments published by Farrell, Bentley and Braski of the Oak Ridge National Laboratory[4]. They irradiated aluminium foils to very high neutron doses, enough to generate ≈ 0.4 at.% silicon by $^{27}Al \rightarrow (n,\gamma) \rightarrow {}^{28}Al \rightarrow {}^{28}Si + e$. When the foils were thinned and examined by scanning transmission electron microscopy (STEM), they were seen to contain not only ordinary voids and silicon precipitates, as expected, but also some anomalous voids sitting on the surface of the foil or extending beyond the foil edge. In fact, these voids, bounded by a dark margin, were inhabiting empty space. Very detailed electron microprobe experments using the ultrafine STEM beam proved that these free-standing voids were coated with shells (4-11 nm thick) of (possibly amorphous) silicon. The aluminium surrounding them had been dissolved away during the thinning process, leaving hollow silicon spheres standing clear of the foils. Silicon has a small solubility in aluminium, and meets Hondros's criterion for ready Gibbsian segregation.

Farrell, Bentley and Braski point out that this substantial Gibbsian segregation offers the hope of influencing the growth rate of voids, especially in their early stages. The segregated solute must alter the specific surface energy at the void walls and thus affect the efficacy of the void in attracting vacan cies — and hence must alter the swelling rate too. If Okamoto and Wiedersich are right in presuming that radiogenic interstitial solutes are peculiarly prone to Gibbsian segregation, then it may be possible to design alloys with initial solutes which, after radiation-induced transmutation, segregate to the incipient voids and inhibit their further growth. That would be an elegant way of setting one consequence of irradiation to neutralise another.

1. J.H. Westbrook, in *Interfaces Conference, Melbourne 1969* (ed. R.C. Gifkins) p. 283 (Butterworth, London, 1969).

2. E.D. Hondros, *ibid.*, p. 77.
3. P.R. Okamoto and H. Wiedersich, *J. Nucl. Mater.* **53**, 336 (1974).
4. K. Farrell, J. Bentley and D.N. Braski, *Scripta metall.* **11**, 243 (1977).

60

Zr_3Al Creeps in a Petty Pace

Nature, **275**, 176 (1978)

In 1956, A. H. Cottrell predicted that metallic uranium, while undergoing steady fission, would lose most of its resistance to creep under stress, and he initiated a very simple experiment with a loaded uranium spring inside a reactor. His hypothesis of enhanced irradiation-creep was promptly confirmed and this had immediate consequences for the design of the British nuclear reactors of that time.

Cottrell's crucial hunch has been followed up in the intervening years by a large body of experiment and theory. Not only a fuel but also its container is subject to this form of creep, and it is a matter of special concern when a 'cladding material' is weakened, for the container has the function of retaining fission products both solid and gaseous; the latter exert a pressure which must be resisted by the container. After much experiment as well as very sophisticated theoretical physics during the past few years, the mechanisms both of irradiation-creep and of the related process of void-swelling (when minute cavities form in the container during reactor service) are now understood in outline. Neutron irradiation produces equal numbers of vacancies and interstitials, some of which recombine quickly; the rest migrate to and become absorbed at 'sinks', of which dislocations are the most important. The crucial point is that a dislocation is a 'biased sink' — that is to say, it has a slightly greater tendency to absorb interstitials than vacancies (the bias is typically around 2%); this bias is explicable in terms of differing elastic interactions of the dislocation with the two types of point defect. The excess of vacancies so generated eventually condenses to form the minute aggregates called *voids* (whose existence was first reported in *Nature* in 1966). Furthermore, an external stress can increase the bias and thus accelerate the climb of dislocations out of their glide plane which must result from a net flux of interstitials to dislocations. That accelerated climb in turn is apt to produce accelerated creep under the action of the external stress, since under certain circumstances, the climb rate is rate-determining for creep.

The recognition of the central importance of point defect kinetics in irradiated alloys has opened the way for a new kind of alloy design — the attempt to predict what kind of alloy might be least susceptible to enhanced creep and void-swelling. The matter is of special interest in fast reactors and in water-moderated reactors; the latter all use container materials based on metallic zirconium, which combines an appropriately low neutron capture cross-section with good corrosion resistance in hot water under irradiation. A determined attempt is now under way in Canada to improve on the 'zircalloys' in current use, and the project is based on a most ingenious hypothesis — the postulated influence of atomic order in an alloy, Zr_3Al, in reducing the net flux of interstitials to dislocations and, thus, in reducing liability to irradiation- creep.

It is well established from comparative studies of successful and unsuccessful industrial innovations that the most important single factor in ensuring success is a resolute and energetic 'product champion'. The champion of Zr_3Al is E. M. Schulson, a metallurgist on the staff of Chalk River Nuclear Laboratories in Canada. [He is now at Dartmouth College, New Hampshire, USA]. He argued in 1972 that this alloy had the virtues of low neutron absorption and promising mechanical properties, and further suggested that irradiation-creep might be reduced. The basic postulate here is that ordering generally reduces the rate of vacancy migration in an alloy (there is an often cited study of beta-brass, in 1956, which first established this point); at any given temperature, a slowing down of vacancy migration gives more time for recombination of vacancies and interstitials and this reduces the net flow of interstitials to biased sinks, and thereby reduces the irradiation-creep rate under a given stress. In pursuit of this hypothesis, Schulson has orchestrated a large range of studies, most of which have come to publication during the past 1½ years: it now looks as though his ingenious hypothesis is justified, though a full study of the irradiation-creep itself is yet to appear.

The hypothesis itself is outlined in a 1977 paper by Schulson[1]. Here he also reports that stress-relaxation of Zr_3Al under neutron or 1 MeV electron irradiation at 570 K (a realistic reactor temperature) is a factor of three or more lower than that for any zirconium-based alloy studied to date. This striking experiment is to be fully described in a conference report to be published by ASTM. Zr_3Al is ordered (Cu_3Au structure, Ll_2 type) up to its peritectoid transformation at 1 265 K, but can be progressively disordered by radiation damage. Several studies have sought to establish whether, at temperatures of practical interest, radiation-disordering goes far enough to nullify the advantage which forms the basis of Schulson's hypothesis. This is a matter of balance between radiation damage and spontaneous reordering, and this balance depends on the rate of damage and the temperature. Accelerated damage by ion bombardment[2] or 1 MeV electron bombardment [3] showed that at realistic service temperatures, large doses leading to several displacements per atom were required to produce substantial dis order above 575 K. Neutron bombardment in reactors would produce less damage and therefore less disorder than either ions or 1 MeV electrons. Further studies[4,5] have established that neutron irradiation strengthens Zr_3Al at all temperatures, and another[6] has elicited the remarkable fact that irradiation actually reduces the notch-sensitivity of the alloy (which is in any case not severe) and, at room temperature, the notched tensile strength is even enhanced by irradiation This behaviour is quite at variance with the normal effect of irradiation on notch- brittleness, which is pronounced, for instance, in ferritic steels. Another study[7] established that the fracture morphology of Zr_3Al indicates a ductie fracture mechanism.

These various experimental studies have now been supplemented by a thorough theoretical investigation of the kinetics of point defect annihilation and irradiation creep for a number of idealised imaginary alloys with various types of atomic order[8]. This very thorough analysis is based on certain assumptions about the change of migration energy for vacancies as a result of ordering and the further assumption that sink strengths are not affected by ordering. Numerical results are obtained for vacancy concentration as a function of temperature, for net interstitial flux to dislocations for various assumed vacancy migration energies and, finally, for the factor of reduction of the rate of dislocation climb due to ordering. In the temperature range of interest, for the idealised Ll_2 alloy, the climb rate (and therefore the irradiation-creep rate) is shown to be reduced by 30--70%, depending on dislocation density and whether simple dislocations or superdislocations are present. This paper marks a milestone in the application of exact atomistic theory to the prediction of irradiation behaviour.

Zr_3Al may turn out to behave even better than theory indicates, because preliminary electron microscopy indicates that the alloy, after irradiation, has a smaller than normal density of dislocation loops. Some unidentified healing mechanism, it seems, may render the alloy even less liable to the processes that enhance irradiation-creep than is predicted by theory.

[1992: It is necessary to add that this promising initiative in the end came to nothing because the alloy's ductility proved insufficient for reactor use. In the history of nuclear power, there have been other prospective cladding materials, such as beryllium, which suffered the same fate.]

1. Schulson, *J. Nucl. Mater.* **66**, 322 (1977).
2. Howe and Rainville, *J. Nucl. Mater.* **68**, 215 (1977).
3. Carpenter and Schulson, *J. Nucl. Mater.* **73**, 180 (1978).
4. Rosinger, *J. Nucl. Mater.* **66**, 193 (1977).
5. Schulson, *Acta metall.* **26**, 1189 (1978).
6. Schulson and Roy, *J. Nucl. Mater.* **71**, 124 (1977).
7. Schulson, *Metall. Trans.* **9A**, 527 (1978).
8. Schulson, Swanson and MacEwen, *Phil. Mag.* **A37**, 575 (1978).

61

Irreversible Segregation in Irradiated Alloys

Nature, **278**, 125 (1979)

The elementary theory of atomic diffusion shows that a solid solution, even if thermodynamically stable, cannot be in a state of complete equilibrium unless the concentration of solute attains a state of flat uniformity. The elementary theory of diffusion is always wrong, because no real solid is structurally uniform. Real pieces of alloy have free surfaces and contain grain boundaries, dislocations and other defects. It is well established that solute segregates to such singularities, and the degree of segregation in equilibrium depends in particular on the solubility of the solute: relatively insoluble species segregate more intensely. The temperature is the other major variable, and a lower temperature spells more segregation. This reversible process is known as equilibrium segregation; it has been intensively studied because of its practical implications, some (by no means all) deleterious. Increased intergranular corrosion and reduced ductility in creep are examples of consequential disadvantages, sintering of 'doped' alumina powder to give a translucent, pore-free solid is an instance of beneficial use. This scientifically most intriguing as well as practically important phenomenon has recently receivcd a worthy review by Hondros and Seah[1]. Equilibrium segregation is not, however, the sole form of segregation in solid solutions.

The alternative form is non-equilibrium or irreversible segregation, and it is associated with flows of point defects, lattice vacancies in particular. Such flows result when a solid is quenched from a high temperature so that it becomes super-saturated with vacancies, or else when it is being continuously irradiated, with the consequential creation of a steady supply of vacancies and interstitial atoms. The flow of excess point defects is absorbed at a 'sink' (such as a free surface, grain boundary or dislocation) and the defects then disappear. If, in some way, the moving defects carry solute atoms along with them, then the resultant segregation is plainly irreversible, because it is brought about by a non-equilibrium, unidirectional process — the flux of point defects.

The study of this kind of segregation has been going on for about ten years and reviews are beginning to appear. An early account was by Anthony, who charmingly entitled it "Atom Currents Generated by Vacancy Winds"[2]; a more comprehensive treatment, including the more recent irradiation studies which only began when Anthony was writing his review, was presented by Harries and Marwick of Harwell at a conference on *Residuals, Additives and Materials Properties* in May 1978 (later published[3] by the Metals Society, London). With the help of some very recent papers, the true complexity as well as the practical implications, for nuclear reactors, of irreversible segregation are now becoming clear, and will be outlined here.

The experimental facts are that under irradiation, solutes (both major and minor) segregate irreversibly to or from a free surface. The experiments are usually done with ion irradiation instead of neutrons, to hasten the process. This was discovered in 1974; a recent study is by Marwick and Piller[4] who found in a Ni-Mn-Si alloy that manganese concentrated in the vacancy-rich layer below the surface (the vacancy source) whereas silicon moved out of this layer towards the sinks; the segregation was very pronounced indeed. The same investigators[5] confirmed the segregation of nickel in a Fe-Ni-Cr alloy to the surface sink.

It has also been found that similar segregation takes place near the surface of the tiny voids that form in some metals and alloys, over a restricted temperature range, during irradiation. (These voids result from the fact that point defect sinks are apt to absorb interstitials slightly faster than vacancies, and so those vacancies which survive both recombinition with interstitials and absorption at sinks, finish up in these minute, empty holes). This segregation process was first found in 1977 in aluminium (see article 59) and has now been confirmed by an exceedingly painstaking electron microprobe study at Harwell[6]. Since the void diameter is mostly in the 100-200 nm range and the irradiation-induced segregation is confined to a layer adjacent to each void of about 100 nm thickness, the utmost resolution of a modern scanning electron microscope was required to demonstrate the segregation, and a number of possible errors had to be checked for. Nevertheless, the findings are quite unambiguous: in the Fe-Ni-Cr-Mo alloy used, the Ni/Fe and Ni/Cr ratios (these being the three major elements) are enhanced near the voids. This matches the earlier finding, that nickel is enriched near surface sinks.

These findings are by no means straightforward to interpret, because the theorist is faced by an embarrassment of alternative transport processes. The most obvious is by the formation of vacancy-solute and interstitial-solute associations; vacancies and self-interstitials in the host lattice become locked to solute atoms with variable binding energies, and the bound solutes are then dragged along with the point defect flux. The most comprehensive attempt to interpret in detail the recent findings in these terms is by Lam, Okamoto and Johnson[7], who improved a model they had been developing for several years. To make sense of the motion of various solutes either with the point defect flux (that is, to sinks) or against that flux (that is, to regions of peak radiation damage), they were forced to postulate very complex patterns of association; for instance, interstitial solute associations were assumed to differ so much in binding strength that some are readily mobile while others cannot move. There are too many adjustable parameters for comfort,

A novel alternative theory has been advanced by Marwick[8]. He concentrates attention on what he terms the "inverse Kirkendall effect". In the normal diffusional Kirkendall effect in a binary solid solution, found in the absence of irradiation, the vacancy flow is governed by the relative partial diffusion rates of the two constituents as they mingle with each other (that is, desegregate). The excess vacancies may form so-called Kirkendall voids near the diffusion interface. Under irradiation, Marwick points out, the situation is stood on its head: a vacancy flow is imposed, and any disparity of partial diffusion rates leads to segregation of solute. The segregation process then creates a countervailing vacancy flux tending to cancel out that generated by irradiation — le Chatelier's Principle in operation. Matwick analyses the process in detail and proves that it is not at all necessary to postulate point defect-solute binding: segregation can result from the inverse Kirkendall effect alone. The faster diffusing species moves up the vacancy gradient (away from sinks) at a rate proportional to the difference between the two diffusivities; the slower diffusing species drifts to the sinks. This model makes sense of the observed enrichment of nickel, a slow diffuser, at surfaces and voids. More generally, Marwick postulates that the inverse Kirkendall effect largely governs the behaviour of concentrated alloy constituents while point defect-solute associations are likely to be the more important feature for dilute solutes.

Much interest has been expressed in the implications of irreversible segregation for void-swelling[8,9]. Structural materials and cladding in fuel assemblies for fast reactors are subject to slight but potentially dan- gerous swelling as voids form in those parts of the assemblies which are set at the relevant temperatures. Intensive research over the past few years has led to the use of cold-worked austenitic nickel-rich alloys (containing Fe+Ni+Cr+Si+Ti) which have excellent swelling resistance at all temperatures. It may now be possible to improve these alloys even further in the light of the new information on segregation, because it is clear (especially if the inverse Kirkendall effect is dominant) that the enrichment of slow-diffusing species, especially nickel, near incipient voids must impede further vacancy access, both because of the preponderance of a species in which vacancies move slowly and also because of the inverse vacancy flux (the balmy vacancy wind) induced by the inverse Kirkendall effect. It seems likely that vacancy access to voids may be choked off almost completely, and some further accurate segregation tests with various solutes may point the way to even more effective anti-void measures. Minor solute elements may have a disproportionate effect if, as is quite possible, their presence has a large effect on the partial diffusivities of the major constituents and thus on the magnitude of the inverse Kirkendall effect. Future improvements may open up the way to fast reactor alloys containing less nickel than the present preferred alloy; nickel soaks up an undue number of neutrons and other alloys might be preferable from the viewpoint of neutron economy.

Another possibility is to prevent void formation altogether by using a non-crystalline alloy in which, presumably, no conventional vacancy flux can arise. Rechtin, Van der Sande and Grant[10] have just reported that glassy $Nb_{40}Ni_{60}$ shows no void formation on heavy ion irradiation, and recommend that glasses like this be explored for use in nuclear reactors.

1. Hondros and Seah, *Int. Metals Reviews* **22**, 262 (1977).
2. Anthony, in *Diffusion in Solids* (ed. Nowick and Burton) (Academic Press, New York, 1975).
3. Harries and Marwick, in *Residuals, Additives and Materials Properties* (ed. A. Kelly *et al.*) 197 (London: The Royal Society, 1980).
4. Marwick and Piller, *Radiation Effects* **33**, 245 (1977).
5. Marwick and Piller, *J. Nucl. Mater.* **71**, 309 (1978).
6. Marwick *et al.*, *Scripta metall.* **12**, 1014 (1978).
7. Lam, Okamoto and Johnson, *J. Nucl. Mater.* **73**, 408 (1978).
8. Marwick, *J. Phys. F: Metal Physics* **8**, 1849 (1978).
9. Harries, *Nuclear Energy* **17**, 301 (1978).
10. Rechtin, VanderSande and Grant, *Scripta metall.* **12**, 639 (1978).

62

The Genesis of a Void Lattice

Nature, **281**, 338 (1979)

When a metal crystal is irradiated with neutrons or other more massive projectiles then, within certain temperature limits, it acquires a population of microcavities known as *voids* : these arise as a result of the preferential agglomeration of the crystal vacancies which are formed together with interstitial atoms as a result of irradiation. In some metals, voids have been found to arrange themselves in an orderly three-dimensional array — a *void lattice.* Properly speaking, this is a superlattice, parallel to but coarser than the underlying crystal lattice. The classic instance is the void lattice in irradiated molybdenum, which mimics the body-centred crystal structure of molybdenum itself: but whereas the interatomic distance in the molybdenum lattice is a fraction of a nanometre, neighbouring voids in the void lattice are separated by a distance of 18-27 nm, according to the temperature at which it is created.

A very extensive and effective research effort has developed during the past decade into the genesis and growth of individual voids in the alloys used to contain nuclear fuels and also into the inhibition of void formation; but the origin of the void lattice, which is of purely fundamental interest, has only intermittently attracted attention. This is a pity, because the phenomenon is intriguing.

A short paper has just been published in which the known facts concerning void lattice formation in molybdenum are assembled and complemented by some further measurements[1]. Molybdenum is the metal in which the existence of a void lattice was originally discovered by Evans in 1971. The new feature which Liou and his collaborators have now established is that voids are originally formed (within a temperature range of about 800-1100°C) always in a random pattern, and a void lattice is formed only later when a much higher radiation dose has been absorbed and individual voids have grown to a sufficient size. Any theory of the formation of the void lattice therefore has to be based upon the progressive ordering of an initially random dispersion of voids, and indeed such a theory must come to grips with the question of why voids have to grow to a certain size before a lattice can form. Notions such as the nucleation of voids on an ordered pattern of dissolved gas atoms or on an ordered array of dislocation loops (such an array has been observed) are not acceptable.

The stability of a void lattice has been satisfactorily explained by Harwell theorists (for example, Stoneham[2]). Spherical voids would attract each other at all separations in an elastically isotropic medium, and it is only in an anisotropic medium that a pair of voids can rest stably at a particular separation. Stoneham has proved that a body-centred cubic or a face-centred cubic lattice of voids is stable in molybdenum, though the predicted ratio of the separation of voids (in the observed b.c.c. void lattice) to the void diameter is smaller than the values actually observed. Stoneham's theory explains why such a lattice is stable once formed, but not how it is established in the first place.

Liou and his collaborators point out that the formation of a void lattice is analogous to the alignment of precipitate plates on different planes to form a preferred, stable array. This is very commonly observed in alloys formed by spinodal decomposition or by ordering, but had not hitherto been properly analysed. The theoretical basis of this phenomenon has now been rigorously treated by Perovic, Purdy and Brown[3], who calculate the total elastic energy of a regular array of plate-shaped precipitates in a matrix, the precipitates lying on three mutually orthogonal planes. They show that such an array, subject to certain conditions, (primarily a low interfacial energy and a large, elastically accommodated misfit between matrix and precipitate), can stabilise at a particular value of the separation of adjacent precipitates, so that further coarsening of the array is inhibited. This presumably is what happens in the formation of a void lattice, though in fact it is not yet known whether a fine void lattice forms first and then coarsens until it becomes stable, or whether the initially disordered voids move about in a kind of Brownian motion until the relative positions of a small group of voids happen to be correct to nucleate a lattice at its proper stable lattice parameter. In the cae of a precipitate array, each precipitate plate creates an anisotropic strain field around itself, and an isotropic medium is assumed, for simplicity; whereas in the case of voids, it is only the anisotropy of the medium which allows a void lattice to form at all.

Electron microscopy has shown that an array of precipitates can coarsen progressively by simultaneous dissolutionof some portions of a precipitate plate and growth of other portions[4]; voids or gas bubbles are known to migrate and merge by a similar mechanism. The growth of voids can be inhibited by small amounts of impurity, which may be fission products such as technetium; the impurity can gather at the void wall and prevent plating out of further vacancies into the void. Because of this, voids may even shrink inthe later stages of irradiation[5]. In such instances, the eventual contraction of the void population is in competition with the formation of a void lattice (which requires a minimum size of void to be attained). What is really required now is simultaneous ion-bombardment and electron microscopic observation of the same samples so that the stages of formation of a void latticecan be observed while they actually happen.

Most published photographs of void lattices show them to be distinctly defective (for instance, the photograph on page 338 of Liou's recent paper); the periodicity has a good deal of scatter. It would be interesting to know whether a void lattice, once formed, gradually loses such regularity as it possesses if it is progressively heated; just as a conventional superlattice in, say, Cu_3Au gradually disorders as the alloy is heated. No experiments on annealing of void lattices seem to have been attempted, nor has anyone attempted to improve the regularity of a void lattice by very slow cooling (which is the method used to obtain highly perfect conventional superlattices). There are plainly a number of parallels between void lattices and conventional superlattices which deserve critical examination.

1. Liou, Smith, Wilkes and Kulcinski, *J. Nucl. Mater.* **83**, 335 (1979).
2. Stoneham, *J. Phys. F* **1**, 778 (1971).
3. Perovic, Purdy and Brown, *Acta metall.* **27**, 1075 (1979).
4. Saunderson, Wilkes and Lorimer, *Acta metall.* **26**, 1357 (1978).
5. Evans, *J. Nucl. Mater.*, in the press [at that time].

63

Novel Model for Cavity Lattices

Nature, **329**, 284 (1987)

Microscopic cavities, either empty (voids) or gas-filled (bubbles), created in crystalline solids by irradiation or ion implantation, are apt to form lattices that mimic the host lattice but are coarser by a factor of 70-120. These entities are termed *void* or *bubble lattices* ; *cavity lattice* covers both categories. Void lattices were originally discovered in molybdenum by Evans[1], at the Atomic Energy Research Establishment at Harwell, England, where much of the subsequent work has been done. Cavity lattices have been observed in metals with various structures, alumina and alkaline earth halides. Johnson *et al.*[2], also working at Harwell, on page 316 of this issue on *Nature* , propose a radically new mechanism, an inter-cavity repulsion caused by 'dislocation punching'. This joins several other mechanisms that have been proposed[3,4] to account for the formation of cavity lattices.

Cavity lattices in cubic metals are always in parallel orientation and fully developed in three dimensions; in hexagonal close-packed (hcp) metals, however, voids aggregate in sheets parallel to the basal plane, without internal ordering in the sheets. Most current theoretical models address the formation of parallel-oriented cavity lattices. It was recognized soon after Evans's original discovery that an elastic interaction between voids or bubbles arises from the anisotropy of the elastic constants; there are both attractive and repulsive components, so that an optimum separation between cavities results. But elastically isotropic metals like tungsten also form voids, a difficulty that is overcome if the spherical voids are allowed to distort and develop flat faces. Although the stability of a lattice, once formed, can certainly be explained in terms of this rigorously developed static model, it is still uncertain whether the energy minimum at the preferred cavity separation is deep enough to permit lattice formation in the first place on the basis of this model (J.H. Evans, personal communication).

More recently, dynamic theories have been developed on the

basis of anisotropic diffusion of self-interstitial atoms (SIAs); in the original form, constrained one-dimensional diffusion or dynamic replacement sequences were assumed. This kind of mechanism can be effective because unaligned voids will receive a greater flux of SIAs than those which are already aligned, since the latter will be locally shadowed by cavities already in the lattice. Cavities migrate in a crystalline solid by a mechanism involving surface self-diffusion along the cavity wall — the more rapidly, the smaller the cavity — and the flux of SIAs accelerates such migration. Local shadowing acts as a feedback control mechanism tending to perfect a lattice. The most recent variant of this approach, originally proposed in 1972, is by Woo and Frank[5]. This kind of model has also been embraced by those seeking to interpret cavity lattice formation in non-metals[4].

An alternative form of SIA model, recently advanced by Evans[6], involves two-dimensional diffusion of SIAs confined within a crystal lattice plane. Computer simulation has shown that this kind of motion, for which there is a sound theoretical basis, can by degrees convert a random cavity array into an ordered one. This is the only model up to now that can explain the planar ordering of cavities in hcp metals. Unlike the elastic anisotropy approach, however, this model is less clear about the magnitude of the interaction holding the lattice in place once it is formed. The older model interprets stability, the newer ones concentrate on initial formation.

The latest Harwell model is intended to interpret a novel set of observations, also reported in the paper by Johnson *et al.* [2]. The authors injected helium into copper (a venerable technique, first used in 1960 by Barnes and Mazey — the latter is one of the authors of the new paper — in the demonstration of bubble mobility in a solid metal). Helium bubbles about 2 nm in diameter are created at a concentration of about 10^{25} m^{-3} — an unusually high concentration — and form a bubble lattice, regular enough to give satellite maxima in electron diffraction. What is novel about this lattice is that parts of it are *not* in parallel orientation: two types of domain are present sharing a common [111] plane with the parallel- oriented domains, but are mutually rotated about the perpendicular <111> direction. This is the first firm observation of cavity lattice domains in non-parallel orientation.

The authors go on to point out that small noble-gas bubbles formed by ion implantation are always grossly overpressurized, in the sense that the surface tension alone does not suffice to contain the gas pressure. The crystal surrounding each bubble is consequently under great stress. Such bubbles can grow by absorbing vacancies created by radiation damage caused by the implanting beam, but when these are exhausted, the only further way a bubble can grow (and thus reduce its overpressure) is either by merging with another bubble — not feasible in a lattice with repulsion between bubbles that approach each other too closely — or by plastically deforming the stressed crystal matrix surrounding the bubble.

One mechanism of plastic deformation that has been well documented[7,8] is dislocation punching: a succession of mobile dislocation rings, of the same diameter as the bubble, is dispatched from the bubble along a 'glide cylinder', the axis of which coincides with the Burgers vector, <011>, away from the surface of the bubble, providing more room for it.

Thus, neighbouring helium bubbles disposed so that a <011> vector joins them, experience a mutual repulsion from the dislocation rings which each bubble emits towards the other, as has indeed been previously recognized[9]. (Dislocations of the same sign necessarily repel each other and their sources.) This mutual repulsion may suffice to overcome the elastic attraction which holds the bubble lattice together, and Johnson *et al.* [2] demonstrate that this process will convert the parallel orientation into one or other of the two observed alternative orientations, in both of which the bubble separations parallel to <011> are increased by a factor of at least $\sqrt{3}$. The experiments reported in this issue involve particularly high helium concentrations and close bubble spacings (about 6 nm) so that bubble repulsions should be unusually strong.

Studies of cavity lattices such as this have a bearing on the possibility of ordering, on lattices, of other structural entities. Khachaturyan and Airapetya[10] have demonstrated that elastic interactions can account for the formation of *precipitate* lattices. In principle, precipitates could also stress the surrounding matrix crystal lattice if their unconstrained atomic volumes exceed that of the matrix lattice, and conceivably dislocation punching could also be involved in the formation of this kind of macrolattice.

1. J.H. Evans, *Nature* **229**, 403 (1971).
2. P.B. Johnson, A.L. Malcolm and D.J. Mazey, *Nature* **328**, 316 (1987).
3. K. Krishan, *Radiation Effects* **66**, 121 (1982).
4. E. Johnson and L.T. Chadderton, *Radiation Effects* **79**, 183 (1983).
5. C.H. Woo and W. Frank, *Mater. Sci. Forum* **15-18**, 875 (1987).
6. J.H. Evans, *Mater. Sci. Forum* **15-18**, 869 (1987).
7. R..S. Barnes and D.J. Mazey, *Acta metall.* **11**, 281 (1983).
8. J.H. Evans, *J. Nucl. Mater.* **76 & 77**, 125 (1978).

9. V.I. Dubinko, V.V. Slezov, A.V. Tur and V.V. Yanovsky, *Radiation Effects* **100**, 85 (1986).

10. A.G. Khachaturyan and V.M. Airapetyan, *phys. stat. sol. (a)* **26**, 61 (1974).

Magnets

64

Ultrahard Magnets

Nature, **230**, 123 (1971)

Four years ago, Strnat and his coworkers discovered that the compound YCo_5 (with the $CaCu_5$ crystal structure) has an extremely high magnetocrystalline anisotropy, and a correspondingly high coercitivity under appropriate conditions. Strnat's discovery stimulated a good deal of urgent research on this and other isomorphous rare earth compounds. Coercivities of 5 000-15 000 oersted can now be obtained by grinding one of the compounds into particles of size <20 μm, and then aligning the powder in wax in an applied magnetic field. Because the anisotropy field is so large (around 300 kilooersted for $SmCo_5$, the most impressive compound), it seems clear that the coercivity cannot be limited by domain rotation, and the nucleation and migration of domain walls in the particles must be postulated.

McCurrie, Carswell, and O'Neill[1] have made a critical comparison of $SmCo_5$ and $LaCo_5$ and have sought in particular to account for the much higher coercivity of particles of the samarium compound. They suggest, following some calculations by Aharoni, that groups of dislocations and stacking faults should serve as easy nucleation centres for reverse domains, and that consequently complete brittleness — implying immobility of dislocations and consequently their inability to multiply — should make for high coercivity. Using a Knoop microhardness tester, they established that the hardness of $LaCo_5$ is highly anisotropic in the $(10\bar{1}0)$ plane though not in the (0001) plane, whereas $SmCo_5$ is isotropic throughout and moreover is distinctly harder overall than $LaCo_5$. From their measurements they deduce that dislocations are mobile on the $(10\bar{1}0)$ plane in $LaCo_5$ but not in $SmCo_5$; no certain deduction is possible concerning dislocation mobility in the basal plane, but the crystal structure suggests that if any slip is possible at all, it should be on $(10\bar{1}0)$. So far as they go, these observations are therefore consistent with the hypothesis that crystal defects reduce coercivity in these materials, and a recent observation that annealing the crushed powder can greatly enhance the coercivity of $SmCo_5$[2] is also in line with the hypothesis. Dislocations, it seems, tend to soften a hard magnet whereas they are well known to harden a soft magnetic material.

Bachmann, Bischofberger and Hofer[3] demonstrate domain patterns in individual $SmCo_5$ particles by using the Kerr effect in conjunction with a polarizing metallurgical microscope. They find cone-shaped reverse domains at the surface of particles larger than about 25 μm, but a quite different domain morphology in small particles with high coercivities. Grain boundaries within a particle were found to be sites where reverse domains can nucleate readily, and indeed polycrystalline powders have low coercivities. One puzzle which the authors hope to resolve by means of their observational technique is the fact that the optimum coercivity in SmC_5 is obtained for a particle size of about 10 μm, which is much larger than the largest size which should energetically prefer to be in the form of single domains. It should be particularly interesting to test McCurrie's ideas about the role of dislocations, and their removal by annealing, by using Bachmann's method of observing domains, and no doubt this will be one of the variables to be examined by Bachmann's team in the systematic study now under way.

1. McCurrie, Carswell and O'Neill, *J. Mater. Sci.* 6, 164 (1971).
2. Westendorp, *Solid State Commun.* 8, 139 (1970).
3. Bachmann, Bischofberger and Hofer, *J. Mater. Sci.* 6, 169 (1971).

65

Wanted – A Taxonomist for Magnetism

Nature, **246**, 445 (1973)

Until quite recently, magnetism was a nice, tidy branch of physics. Every student of the field knew that it was neatly divided into diamagnetism, paramagnetism, ferromagnetism, ferrimagnetism and antiferromagnetism: Néel was in his heaven and all was right with the world. Such a sense of tidiness is apt to precede a thoroughgoing shake-up, and magnetism has been no exception. Now there is mictomagnetism, (crystalline) superparamagnetism, amorphous antiferromagnetism and, as the most recent candidate, speromagnetism. What this amounts to is that magnetism has come face to face with the physics of randomness, and randomness has turned out to be more subtle than mere regularity.

Speromagnetism is a term coined by Coey of Grenoble and Readman of Edinburgh in their letter on page 476 of this issue of *Nature*. They report on an examination of a natural gel consisting of hydrated amorphous ferric hydroxide in an organic matrix. The magnetization curve at low temperatures shows a remanence, which indicates that there must be some kind of ordered magnetic structure. Combining magnetization studies with a detailed examination by Mössbauer spectroscopy, the authors conclude that below 10 K the spins are coupled and locked within small particles of the order of 10 nm in size; above 10 K, these particles become 'unblocked' and behave superparamagnetically — that is, the particles are now magnetically independent but in each the spins are still fixed in the absence of a constraining field. Only above 100 K does the material become a self-respecting paramagnet, with thermally randomised spins.

Applying Langevin theory to the superparamagnetic magnetization curve leads to a net moment of 73 μ_B per particle, and also to an estimate of the particle size as equivalent to 635 formula units of the hydrated ferric hydroxide. This at once poses a paradox, because 73 μ_B for 635 formula units is consistent neither with ferromagnetic nor with antiferromagnetic coupling within a particle. The authors are driven to the conclusion that the spins are arranged effectively at random. The point to notice is that a purely random array of N spins of moment m in an independent small particle has a resultant moment of $\sqrt{N}m$: this is simply the random walk equation, familiar from diffusion theory. Although the authors do not trouble to spell it out, the resultant saturation magnetization is much greater for a population of spins distributed in many minute particles than for the same population arranged in a few large particles.

The authors call the behaviour of an array of small particles with fixed but random spins, which have an appreciable resultant moment only because of the small size of the array, by the name 'speromagnetism', from a Greek root meaning 'to scatter'. The interesting question is why an array of ferric spins is arranged randomly in spite of the undoubted presence of some antiferromagnetic nearest-neighbour interactions, and on this they are able only to speculate. If seems that the existence of variable exchange interactions in an irregular structure must in some way be held responsible.

This intriguing paper links with a substantial body of recent research concerned with the magnetic properties of extensive (rather than particulate) amorphous materials, dignified this year with a full-scale volume of conference proceedings[1]. Both ferromagnetic and (even more surprisingly) antiferromagnetic characteristics have been observed in alloy glasses. The theory of spin assemblies where the strength and even the sign of exchange interaction vary according to the spacing of randomly packed neighbours is very complex and has only recently been attempted. The new phenomenon of speromagnetism (or amorphous superparamagnetism) adds a further delectable layer of complexity to what is already a highly elaborate network of phenomena. Truly the services of a magnetic taxonomist will soon be needed to bring order out of the amorphous multiplicity of concepts.

1. H.O. Hooper and A.M. de Graaf (editors), *Amorphous Magnetism* (Plenum, New York , 1973).

66

Searching for Supermagnets

Nature, **317**, 112 (1985)

Lodestones have long been superseded by increasingly esoteric compounds in the search for magnetic power. The latest novelty is a class of Fe-Nd-B compounds which form the most powerful permanent magnets yet created. The performance of these is based on the tetragonal compound $Nd_2Fe_{14}B$ which, since its identification in late 1983, has been the subject of over a hundred papers. There has never before been a comparable worldwide burst of effort to produce, understand and improve a new magnetic material.

As described by Robinson[1], the starting point was work done in 1973 by Arthur Clark of the US Navy. The issue was how to improve on the then best available magnets, Sm_5Co and Sm_2Co_{17}. The basis of such magnets is that the rare-earth metal provides high crystal anisotropy and therefore high coercivity, while the transition metal raises the Curie temperature high enough to avoid restrictions on use. Attempts to replace cobalt with iron in the hope of enhancing the magnetization without sacrificing coercivity were generally unimpressive, but RFe_2 (where R =Tb, Dy or Sm) compounds were much better, provided they were prepared by making and then crystallizing a glassy structure. Evaporation was in due course replaced by melt-spinning a jet of molten metal against a spinning copper wheel, using boron to stabilize a glassy structure. By the beginning of 1983, Fe-Tb-La-B and Fe-Nb-B ribbons had been prepared exhibiting high coercivity — the source of which was a mystery. Then it was discovered that the X-ray diffraction patterns from melt-spun and annealed Fe-Pr-B-Si ribbon showed a previously unknown phase, identical with one which had heen observed, but not analysed, in 1979 in the Fe-Nd-B system; within a few months this phase was identified as $Nd_2Fe_{14}B$ and its structure determined in three laboratories[2-4].

In June 1983, the Japanese company Sumitomo announced the production of Fe-Nd-B permanent magnets by the straightforward sintering route which works so well for the Co-Sm magnets, though for the new material the heat-treatment after sintering is critical. The compositions of both the sintered and the melt-spun versions of Fe-Nd-B have to be displaced from $Nd_2Fe_{14}B$ stoichiometry for best magnetic properties, though the needful displacement is different for the two production routes. To summarize a great deal of research published since mid-1983, the compositional displacements seem to be needed to permit other, ultrafine, phases, both magnetic (α-Fe) and non-magnetic (Fe_4NdB_4 and a Nd-rich phase) to precipitate at the boundaries of the $Nd_2Fe_{14}B$ grains. This adds the effects of domain-wall pinning to the large magnetocrystalline anisotropy of the main phase.

For sintered magnets, domain-wall pinning is certainly crucial, since the grain size (5-15 μm) is much larger than the 0.3 μm diameter calculated as the 'single-domain particle diameter', D_c. However, for the melt-spun material (which, it seems, always passes through the glassy phase at some point in the process) the grain size, controlled by the quench rate, that gives optimum properties is 20-80 nm, considerably smaller than D_c. But the recognized experts on the rapid-quenching route at General Motors are convinced that the optimum grain size is in some way connected with the achievement of single-domain morphology, in spite of the mismatch with D_c. The relation between microstructure and magnetic properties was thoroughly analysed in a very up-to-date review by J.D. Livingston[5] presented at a recent meeting on rare-earth magnets, and also in a paper by Hadjipanayis and colleagues[6], first presented at last November's '3M' conference, at which the Japanese also gave an account of their sintered Fe-Nd-B material[7].

Since their original account[8] in early 1984, the General Motors team have said little, but very recently Lee has published an account of the last year's developments[9]. One of the weaknesses of the melt-spun material (whether dispersed in a non-magnetic matrix or hot-pressed) has been its magnetic isotropy. Now, by die-upsetting (a process in which transverse plastic deformation is allowed) following hot-pressing, strongly anisotropic material can be prepared, with a much enhanced level of residual magnetization. The fact that $Nd_2Fe_{14}B$ turns out to be plastic at 700°C is one of the many surprising features of this compound.

European physicists and metallurgists have not been left out of this new field, on the evidence of the proceedings, only recently available, of a one-day workshop meeting held in Brussels on 25 October 1984. The influence of economics and the supply position is particularly clearly set out in this report — the relative scarcity of both cobalt and samarium stimulated the search for alternatives to Co-Sm magnets— and it is clear that neodymium, the most plentiful of the 'rare' earths, is not rare at all, being more abundant than lead. Perversely, the most serious practical difficulty with $N_2Fe_{14}B$ — the low Curie temperature, T_c — can be readily overcome by adding cobalt . Of course, not only does this imply a renewed dependence on scarce cobalt supplies, but it also lowers the saturation magnetization. An intriguing alternative[10,11] that may have industrial potential is to raise T_c and magnetization by absorbing hydrogen into $Nd_2Fe_{14}B$; Fruchart *et al.* [11] report a T_c increase of 87 K.

Several papers presented at Brussels or Dayton dealt with possible applications of Fe-Nd-B based magnets, both within and outside the motor industry. The size of the research programme mounted by General Motors is sufficient indication of the potential for the use of high-grade permanent magnets, and these remarkable new materials offer scope both for fundamental research into the causes of magnetism and for profitable development.

1. A.L. Robinson, *Science* **233**, 920 (1984).
2. J.F. Herbst, J.J. Croat, F.E. Pinkerton and W.B. Yelon, *Phys. Rev. B* **29**, 4176 (1984); *J. Appl. Phys.* **57**, 4086 (1985).
3. D. Givord, H.S. Li and J.M. Moreau, *Solid State Commun.* **50**, 497 (1984)
4. C.B. Shoemaker, D.P. Shoemaker and R. Fruchart, *Acta Cryts. C* **40**, 1665 (1984).
5. J.D. Livingston, in *Proc. 8th Int. Workshop on Rare-Earth Magnets and Their Applications* , p. 423 (University of Dayton, 1985).
6. G.C. Hadjipanayis, K.P. Lawless and R.C. Dickerson, *J. Appl. Phys.* **44**, 148 (1984).
7. M. Sagawa, S. Fujimura, H. Yamamoto, Y. Matsuura and S. Hirosawa, *J. Appl. Phys.* **57**, 4094 (1985).
8. J.J. Croat, J.F. Herbst, R.W. Lee and F.E. Pinkerton, *Appl. Phys. Lett.* **44**, 148 (1984).
9. R.W. Lee, *Appl. Phys. Lett.* **46**, 790 (1985).
10. J.M.D. Coey, J.M. Cadogan and D.H. Ryan, in *Proc. Nd-Fe Permanent Magnets – Their Present and Future Applications* (I.V. Mitchell, ed.) p. 143 (Commission of the European Communities, Brussels, 1984).
11. S. Fruchart *et al.* , as ref. 10, p. 173.

67

Concerted European Action on Magnets

Advanced Materials, **2**, 518 (1990)

In 1967, the sedate world of permanent magnets was set agog by the report that the Compound $SmCo_5$, crystallizine in the familiar $CaCu_5$ structure type, was a far stronger magnet than the best alloy magnets then current. $SmCo_5$ was soon joined by Sm_2Co_{17}. The perfomance of a permanent magnet material can be measured in terms of the intrinsic coercive 'force', or field, though a better practical measure is the maximum energy product (the largest value of the product of magnetization and energizing field selected those measured at all points around the magnetic hysteresis loop). Figure 1 shows how these values shot up in the late 1960s and subsequently. A good, concise account of the history of permanent magnet materials since 1880, culminating with rare-earth magnets, has just been published by Livingston[1].

The Sm-Co magnets made an excellent industrial start, but then disturbances in the Congo, by far the world's largest supplier of cobalt, led to such fluctuations in the supply and price of cobalt that the magnetic industry took fright and held back. (It was ironic that the problem arose not with the rare-earth metal but with its supposedly less scarce companion metal). This event led the Commission of the European Communities, from 1982 on, to include permanent magnet materials within the purview of their wider research programme on Materials Substitution, in the hope of promoting the discovery of a 'non-strategic' alternative to cobalt compounds. In 1983, as if in response to this felt need, a new magnetic compound based on neodymium, iron and boron was announced, almost simultaneously in the USA and Japan. (The two companies concerned, General Motors and Sumitomo, are still locked in a patent dispute). Neodymium, more abundant than lead, is the least rare of the 'rare' earth metals. Within a year, four groups had determined the hitherto unknown crystal structure of the alloy's 'active principle,' the compound $Nd_2Fe_{14}B$. The almost fortuitous circumstances of the discovery of this phase and its early history were mapped out in *Nature* in 1985[2]. An excellent critical survey of intermetallic compound permanent magnets, with heavy emphasis on $Nd_2Fe_{14}B$, appeared the following year[3]: it is remarkable how much had already been established by that time.

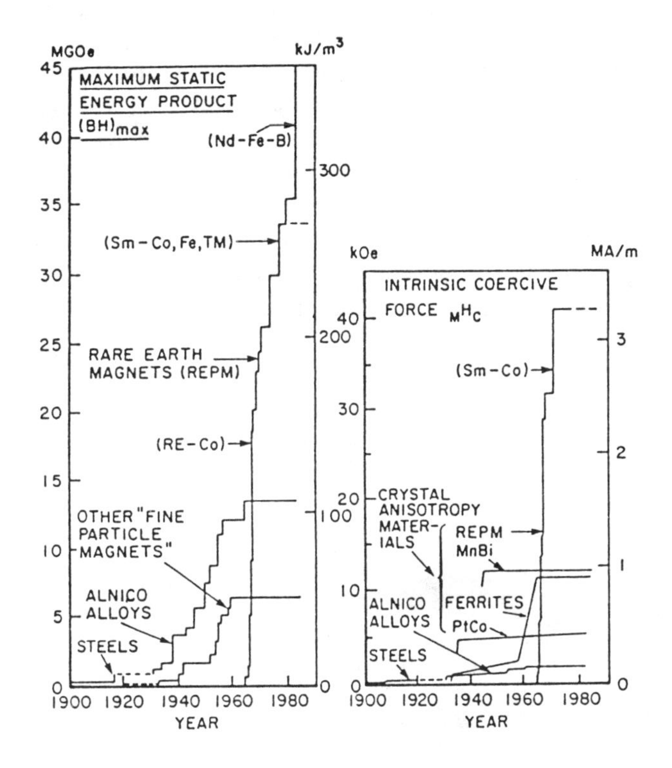

Fig. 1 Permanent magnet development since 1900, showing values of the maximum static energy product (left) and the intrinsic coercive field.

The European Commission, on the urging of several European organizations, arranged a first European Workshop on Nd-Fe-B in Brussels, in October 1984, and quickly published the proceedings[4]. This seminal meeting then led to something quite novel: with strong encouragement from the Commission's Science Directorate, a group of scientists and engineers from a wide spectrum of organizations drew up a proposal for a closely concerted research programme, and in February 1985, CODEST, the Community's advisory group for science and technology, approved a proposal for funding from the aptly named *Stimulation* programme, to the tune of 2.5M ECU. (It is ironic that recently, on the grounds of possible offensive misinterpretation by those not wholly au fait with the subtleties of the English language, the programme's name was changed from*Stimulation* to *Science*).

The *Concerted European Action on Magnets* or CEAM (whose logo appears here) was innovative both in its objectives and in its organization. The contract was placed exclusively with four individuals in four countries: Prof. Givord of the Louis Néel magnetics laboratory of the CNRS in Grenoble, France; Prof. Coey of Trinity College, Dublin, Ireland; Prof. Harris of the University of Birmingham, England; and Prof. Hanitsch of the Technical University of Berlin, Germany. Other academic and industrial laboratories, 58 in number in due course, were subcontracted by these four. The work to be done was subdivided into three areas: Materials, under the charge of Givord and Coey, physicists; Magnet Processing under Harris, a metallurgist; and Applications, under Hanitsch, an electrical engineer. Each leader had about 15 laboratories to coordinate.

Fig. 2. The logo of the Concerted European Action on Magnets (CEAM) programme.

A crucial feature of the master contract was that the funds were to be spent largely on purposes other than salaries and hardware; they went to pay for (1) frequent project conferences, (2) extensive exchange of research workers, (3) the publication and distribution of a newsletter dedicated to rapid information exchange between participants, and (4) the creation of a computerized database on novel permanent magnet materials and their uses. The specific costs of the research itself were mostly met by the participating organizations themselves, by national funding bodies or by other European Community contracts financed from the European Advanced Materials (EURAM) programme.

CEAM was a striking success, so much so that its original two-year duration was prolonged for a third year into 1988, and then in 1989 it was succeeded by a follow-up pro-gramme, CEAM II, wholly targeted on the four functions listed above. CEAM I involved, over its three years, some 180 scientists and engineers (including students); about a third of these spent periods in participating laboratories other than their own; 356 papers were published and 8 patents filed as a direct result of CEAM I. 63 of these papers appeared in the proceedings[5] of a Conference held in Madrid in April 1988, which constituted a grand finale to CEAM I. An indication that one of the special objectives of CEAM I had been achieved is the fact that a third of the 356 published papers involved authors in two or more laboratories, divided among eight Community countries plus Austria.

The proceedings of the 1988 conference open with four overview papers: one, by Mitchell[6], reviews the organization in which he himself played a vital energizing part from CEC headquarters, collaborating closely with the first scientific head of the enterprise, Pauthenet of Grenoble (after whose early death Coey took over). The next paper[7] by the four remaining editors of the proceedings, provides an effective overview of the scientific results obtained by CEAM I.

The *Materials Activity* involved mostly physicists and chemists who examined phase diagrams, searching for and comparing new alloy phases. (Indeed, a number of promising new compounds and alloys, such as for instance $Sm_{20}Fe_{70}Ti_{10}$[8] have already been identified). The participants studied the microstructure of Nd-Fe-B magnets and related this to measured magnetic parameters, and further did some fundamental work on recalcitrant and neglected issues such as the mechanisms that control the nucleation of reverse magnetic domains (which often determines the coercive field). This range of research, which involved as many laboratories as the other two Activities combined, led to the lion's share of the papers and provided most of the patents.

The *Magnet Processing Activity* involved metallurgists and materials scientists and much collaboration between universities and industry. The research here was firmly focused on Nd-Fe-B and included a detailed examination of an entirely new processing approach — hydrogen decrepitation, in which a coarse powder is hydrogenated, which breaks it down into a fine powder ideally suited to sintering, followed by pumping off the hydrogen. A critical comparison of the two established approaches for making Nd-Fe-B magnets, sintering and rapid solidification, was also performed. Two industrial producers were subcontracted to prepare ingots of various Nd-Fe-B compositions for the entire CEAM I project, so that results from different laboratories could be usefully compared. Corrosion protection of Nd-Fe-B magnets, which proves to be important for their long-term application, was also studied in detail.

The third Activity, concerned with *Applications* , was in some ways particularly impressive. There was heavy emphasis in the participating laboratories (industry, of course, played a substantial part) on computer-aided design to exploit powerful magnets to best advantage. Nd-Fe-B magnets can be very compact indeed while still providing the necessary flux and resisting demagnetization, and optimizing a design requires many competing factors to be taken into account. The impression is left that the promise of these magnets is greatest for small brushless motors (of which, for instance, dozens can be used in a single automobile[1]...hence General Motors' massive investment in manufacturing Nd-Fe-B alloys), in stepping motors, and also in applications requiring an extremely uniform field: thus a complex magnet assembly was designed for use in a medical NMR body scanner. Designs were also worked out for 'solenoid substitutes' based exclusively on Nd-Fe-B permanent magnets in which, by rotating concentric component tubular magnets relative to each other, an axial field strength can be continuously varied. Another project was devoted to designing handling devices to allow the small but very powerful magnets in, say, a stepping motor to be safely assembled — not, it turns out, at all a simple matter.

In two issues (February and April 1989) of the CEAM newsletter, Coey presented the concept, not at that time made actual, of a "magic magnet" in which the flux density in an air gap can considerably exceed the remanence, and Spooner suggested some applications. The concept rests on the extremely high magnetocrystalline anisotropy of the alloy, which allows differently oriented parts of a segmented magnet to retain a preferred magnetization direction, hardly affected by continuous sectors. This allows an unprecedented degree of freedom in designing flux vectors in the solid for optimum perfomance.

The effective coordination of the fundamental and applied sectors of CEAM represents something all too rare in magnetism. In 1985, the American magnetician White reviewed[9] the research opportunities in magnetic materials and pointed out that "Magnetism... has both a basic as well as an applied element; ...the gulf between these two areas is very deep. As a result, physicists are often not aware of the fundamental problems limiting the technology, and the applied scientists are not aware of new solid-state techniques that could provide answers to their problems. ...Simply bringing these two cultures closer together would stimulate interactions that would benefit the technology". This is what CEAM has effectively achieved.

1. J.D. Livingston, *JOM* (formerly *Journal of Metals*) **42**, 30 (Feb. 1990).
2. R.W. Cahn, article 66, above.
3. K.H.J. Buschow, *Mater. Sci. Repts.* **1**, 1 (1986).
4. I.V. Mitchell (ed.), *Nd-Fe Permanent Magnets: Their Present and Future Applications* (Elsevier Applied Science Publishers, London, 1985).
5. I.V. Mitchell, J.M.D. Coey, D. Givord, I.R. Harris and R. Hanitsch (eds.), *Concerted European Action on Magnets (CEAM)* (Elsevier Applied Science, London, 1989).
6. I.V. Mitchell, in ref. 5, p. 1.
7. J.M.D. Coey, D. Givord, I.R. Harris and R. Hanitsch, in ref. 5, p. 17.
8. K. Schnitzke, L. Schultz, J. Wecker and M. Katter, *Appl. Phys. Lett.* **54** , 587 (1990).
9. R.M. White, *J. Appl. Phys.* **57**, 2996 (1985).

Superconductors and Superionic Conductors

68

Superionic Conductors

Nature, 235, 133 (1972)

Anxieties about the ill-effects of traffic fumes and the problems of maintaining oil supplies combine to keep alive interest in the potential of electric cars. Such cars require either fuel cells (currently in the doldrums) or rechargeable batteries, and these must be designed for high energy density. This condition can best be met, it currently appears, by high-temperature batteries, of which Ford's sodium-sulphur battery is the best known, and such cells need electrolytes capable of standing up to fierce chemical onslaughts at high temperatures.

The sodium-sulphur cell (which is currently being examined in a number of industrial laboratories, including British ones) makes use of a solid electrolyte, beta-alumina. This extraordinary material is a refractory which, in spite of its name, contains some sodium and approximates to the composition $Na0.11Al_2O_3$; the sodium (or any of a number of alternative metals) stabilizes the anomalous crystal structure. Beta-alumina has an extremely high conductivity, much closer to that of a metal than to that of a normal ionic conductor such as NaCl, but, unlike a metal, its conductivity rises steeply with temperature.

Because of its potentially major industrial importance, and simultaneously because it poses a fascinating scientific puzzle, beta-alumina is the subject of much basic-cum-strategic-cum-applied research (pace Rothschild), and a good deal of this has been centred on its crystallographic structure. The most impressive piece of crystallographic work yet to be done on the compound has just been published by W. L. Roth[1]. Roth used small monocrystals in which he substituted silver for sodium by means of ion exchange in a salt bath; he used this method because heavy atoms are easier to locate by X-ray diffraction. The weight change during ion exchange served to establish accurately the sodium content of the original material, which is about 25 per cent above the $NaO.11Al_2O_3$ stoichiometry.

Starting from X-ray diffraction data, a very long and scrupulous process of structural refinement was performed, with the following principal results. All the silver is in the horizontal mirror planes (in other words, planes normal to the hexagonal axis) between the 'spinel blocks' of the structure. There are two kinds of ideal silver sites within the mirror planes, at intersections with the three-fold axes of the space group: the silver atoms are statistically distributed among (and also slightly displaced from) these ideal sites, the rest remaining vacant; the excess charge is probably compensated by interstitial oxygen. There is a huge anisotropy of thermal vibrations of the silver ions, the amplitude being about 1 000 times greater in the mirror plane than normal to it. The structure in the mirror planes has quasi-liquid properties, the silver ions jumping very easily from one site to an adjacent vacant one. (In fact, the r.m.s. vibrational amplitude at room temperature of the silver ions in the mirror plane is almost one tenth of the corresponding interatomic distance which, according to Lindemann's law of melting, is an approxima te criterion for the melting of a solid.) The notion that the conducting ions in compounds of the beta-alumina family move in a two-dimensional quasi-liquid environ ment, confined by blocks of non-conducting ions, is not new, but Roth has unusually clear evidence to substantiate it.

Finally, Roth shows that the apparent randomness of the silver distribution and the hyperstoichiometric silver content can be interpreted on the basis of a model of small domains, only 5-10 Å wide in the *a* direction, in any one of which all silver atoms are restricted to one of the two types of silver sites; the hyperstoichiometry results from the necessary bunching of silver ions in the domain boundaries.

The results require some modification before they are applied to normal (that is, sodium-stabilized) beta-alumina, because there is evidence that the sodium-silver substitution shifts the sites of the metal ions somewhat. Nevertheless, there is now available the basis of a much closer understanding of the essentially two-dimensional superionic conduction of normal beta-alumina and its variants.

The new term, *superionic conduction*, has been introduced by M. J. Rice and W. L. Roth in a survey report which is to be published in the next issue of the *Journal of Solid-State Chemistry* . The term is to apply collectively to three categories of highly conducting ionic compounds: the beta-aluminas, certain silver halides and related compounds among which $RbAg_4I_5$ is the most remarkable), and some "defect-stabilized ceramics", of which calcia-stabilized zirconia is the most familiar. In their article, Rice and Roth replace the standard hopping model of solid ionic conduction by the notion that the conducting ions, given a specific minimum of energy, are freed from their localized states to move as free ions with measurable mean lifetimes and mean free paths (which may be of the order of interatomic spacings or substantially greater, as in calcia-stabilized zirconia). On the basis of this metal-like behaviour of the ions, the authors are able to make predictions about the thermionic power and frequency dependence of conductivity which allow them to substantiate the basic applicability of their model.

1. W.L. Roth, *J. Solid-State Chem.* **4**, 60 (1972).

69

Irrepressible Superconductor

Nature, **269**, 198 (1977)

Those who study superconductors with practical ends in mind have been much concerned to raise the transition temperature and the critical magnetic field: indeed, one sometimes has the impression that a transition temperature at or above ambient represents the philosopher's stone for some of today's physicists. There is however another, less regarded material property which limits potential applications, and that is the critical current density, itself a function of field and a structure-sensitive property. Nb_3Al is one of the most promising A-15 superconductors because of its extremely high critical field (H_{c2} =295 kG at 4.2K), but the critical current density is modest, and various metallurgical tricks used by a number of investigators to modify the microstructure have in the past not been very effective in improving the situation. About 10^5 A cm^{-2} is the best that has been achieved. A completely new approach now promises the engineer an excellent new superconductor.

Lo, Bevk and Turnbull[1] of Harvard University decided to melt-spin the alloy. 'Melt-spinning' is the name given to a variant of splat-quenching which produces a thin continuous ribbon of alloy. The version used here involves a rapidly rotating copper saucer — the substrate — on to which a fine stream of molten alloy is projected. The investigators found that a small silicon addition was needed to produce a good ribbon: their alloy was of composition $Nb_{73}Al_{25.5}Si_{1.5}$. Hitherto, melt-spun ribbons have been used to examine mechanical or ferromagnetic behaviour: melt-spun superconductors are something new under the Sun.

The thickness of the ribbon and thus the effective quenching rate can be modified by altering the speed of rotation of the substrate: the thinnest ribbon was 25 µm in thickness and had grains <0.5 µm in diameter. The microstructure consisted of fragments of dendrites surrounded by transition layers presumed to be of variable composition. The crystal structure remained single phase A-15. To measure electrical characteristics, the samples had to be embedded in indium in order to provide good electrical and thermal connections. Even so, the contact resistance limited the feasible current density to 1.8×10^6 A cm^{-2}. Ribbons 25 μm thick were tested in fields up to 150 kG and the critical current density was found to be in excess of the experimentally enforced limit of 1.8×10^6 A cm^{-2} in all field strengths. When the ribbon was annealed at 750°C, T_c was raised from ≈ 16K to 18.4 K and the critical current density was still in excess of the experimental limit. (It is therefore not known whether heat treatment increased or decreased the critical current density). The highest critical current density in 105 kG for Nb_3Al (that is, the highest in any sub stance) reported in the past was $\approx 10^5$ A cm^{-2}. Plainly an extreme refinernent of microstructure is the key to an enhanced critical current density.

It is already clear from these preliminary results with just one of the A-15 family of compounds. that a new generation of high-field superconductors is in the offing. There will of course be problems — principally the need to incorporate superconducting ribbons into a thermally well-conducting matrix such as copper to provide insurance against a breakdown of superconductivity in service. A number of metallurgical techniques are available for this purpose (for example, Tsuei[2]; Chen and Tsuei[3]) but they generally depend on a precipitation of superconducting filamentsi *n situ*. Melt-spun ribbons will have to be incorporated in a matrix after they have been made: it may be possible to use a sintering technique with a low melting metal such as aluminium. The fact that annealing does not damage the superconducting properties will simplify the task.

Rapid quenching from the melt has been used before in connection with superconduction, but in a different sense. Instead of using the technique to refine the structure of a stable compound, as the Harvard scientists have done, workers at California Institute of Technology used it to create metastable structures. Johnson and Poon[4] tested a whole range of metastable simple cubic intermetallic superconductors (such as $Te_{70}Au_{30}$, $Sb_{75}In_{25}$) and established that T_c was systematically (and inversely) related to lattice parameter; $Sb_{75}Au_{25}$ had the highest T_c of ≈ 6.5 K, and it seemed that the (meta) stability of the simple cubic lattice (one atom per unit cell) broke down for smaller lattices. Glassy alloys, made by melt-spinning, can also be superconducting: Johnson, Poon and Duwez[5] examined the superconductivity of Au_xLa_{100-x}, glasses ($16 \leq x \leq 40$). Among other peculiarities, these glasses have ultra- short electronic mean free paths. For $Au_{22}La_{78}$, this path is 1.7 Å, $T_c = 3.4$ K, the upper critical field ≈ 10 kG. This category of material is a scientific curiosity but is not likely to threaten the technological superiority of Nb_3Al and its analogues.

1. K. Lo, J. Bevk and D. Turnbull, *J. Appl. Phys.* **48**, 2957 (1977).
2. C.C. Tsuei, *J. Appl. Physics* **45**, 1385 (1974).
3. Y. Chen and C.C.Tsuei, *J. Appl. Phys.* **47**, 715 (1976).
4. W.L. Johnson and S.J. Poon, *J. Appl. Phys.* **45**, 3683 (1974).
5. W.L. Johnson, S.J. Poon and P. Duwez, *Phys. Rev. B* **11**, 150 (1975).

70

Making A1 A15s

Advanced Materials **2**, 43 (1990)

High-temperature ceramic superconductors receive so much media attention at present that readers might be forgiven for supposing that nobody pays attention to old-fashioned, intermetallic compound superconductors any longer. That would be quite a mistake: the A15 intermetallics remain big business, and superconducting magnets, widely used in research and industry, depend entirely on A15s, notably Nb_3Sn. A15s are of course only a small corner of the rapidly growing research field of intermetallic compounds[1] and many of these compounds are prepared, for research purposes, by rapid solidification[2]. This practice is now so widespread that a recent review article was dedicated to rapid solidification and its application to intermetallic compounds[3]. One category of intermetallic compounds, at least, is now being manufactured for industrial use by melt-spinning, a form of rapid solidification processing: these are the $Nd_2Fe_{14}B$ permanent magnets, made by General Motors in America by melt-spinning to form a metallic glass which is then crystallized to form a very fine-grained material with optimum magnetic properties[4]. These can then be consolidated into large pieces.

Rapid solidification, or quenching from the melt (which means, in practice, a cooling rate of $10^5 - 10^6$ deg/s) has been applied to A15s also, particularly to the experimental production of a metastable A15 phase, Nb_3Si, which can only be made by rapid quenching, either from the vapour or from the melt. Some years ago, Mireille Clapp at the University of Massachusetts, Arnherst, and her collaborator Gresock successfully tried the melt-spinning route for making the A15, superconducting form of Nb_3Si[5]. In this technique, a fine jet of molten alloy is projected by gas pressure against a rapidly spinning polished copper wheel. Gresock and Clapp developed an unusual melt-spinning apparatus, described in detail in this early paper, in which the reactive alloy is suspended out of contact with the crucible and only drops on to the ejection nozzle just before it is ejected; one consequence of this arrangement, which required much fine-tuning, is that the melt is not superheated much above the melting-point, important when that melting-point is around 2000°C. With a very short contact time and avoidance of superheating, the silica crucible survived the impact of the very hot melt. The as-solidified ribbon was amorphous and was then heat-treated to crystallize it into the fine-grained A15 compound.

Armed with this first success, Prof. Clapp's team then went on to melt-spin other A15 compounds, at first two unconventional compositions, $Ti_3Nb_6Mo_3Si_4$[6] and $Nb_{73}Al_{12}Si_{14.5}B_{0.5}$[7] and most recently the more familiar Nb_3Ge (with 0.5 % of B added)[8]. The objective of the added boron and the complex compositions of the first two alloys was to ease the production of a metallic glass, which is always aided by the presence of metalloids such as B and Si and by multiple solutes. (In the latest paper, some improvements to the melt-spinning procedure are also described). In each instance, the resulting glass was then crystallized. This sequence of experiments had been leading up to the synthesis of Nb_3Ge, which has particularly good superconducting parameters but is difficult to make by conventional routes, and the 1989 paper reported excellent critical currents of 5×10^{10} and 8×10^{10} A m^{-2} at fields of 0 and 15T, respectively. The authors point out that their approach could be a stage in the production of multifilamentary Nb_3Ge superconducting tapes or wires.

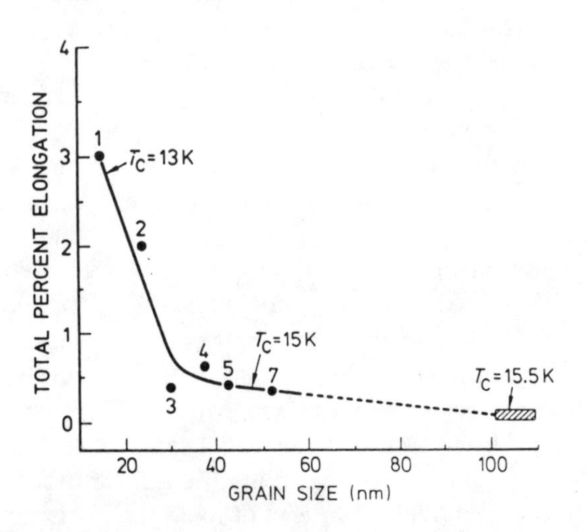

Fig. 1. Effect of grain size on the total percent elongation of A15 $Nb_{73}Al_{12}Si_{14.5}B_{0.5}$ tapes. (1) 24h, 629°C. (2) 48h, 620°C. (3) 48 h, 660°C. (4) 10h, 700°C. (5) 24h, 700°C. (6) 20h, 750°C. • Amorphous, annealed. Shaded area, liquid-quenched. (After ref. 7).

In the course of these experiments, it was found that the crystallized glasses were apt to have extremely fine grain sizes. Different heat-treatments of the glassy ribbons generated varying grain sizes. Clapp and Shi[7] found that when the grain size was reduced to below 30 nm (compared with a normal grain size of several μm), the normally entirely brittle compound acquired some capacity for plastic deformation, as evinced in simple bending of the meltspun ribbons. Figure 1 shows their results with meltspun A15 $Nb_{73}Al_{12}Si_{14.5}B_{0.5}$ tapes: a plastic surface strain of 3% without fracture is something quite remarkable in materials of this kind. The scope of very fine grain or subgrain sizes for overcoming brittleness has not been widely investigated, but it is readily enough understood, since arrays of dislocations piling up against a grain boundary magnify the externally applied stress so as to nucleate a crack at the boundary. The smaller the grain, the smaller and the less damaging the dislocation pileup. The literature concerning ductilization by ultrafine grain structures has been reviewed by Cahn[9]. It is not clear whether the plastic deformability indicated in Figure 1 is due to dislocation motion or to diffusional creep involving the motion of point defects across the fine grains: a study of the stress dependence of flow rate should resolve this uncertainty. Since brittleness in A15 compounds is a serious problem in the shaping or superconducting coils for magnets, Clapp and Shi's finding is likely to be of distinct practical importance.

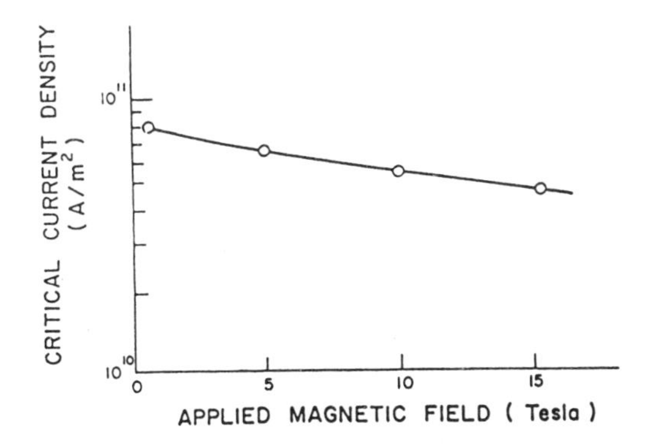

Fig. 2. Critical current density vs applied magnetic field at 4.2 K for a $Nb_{73}Al_{12}Si_{14.5}B_{0.5}$ sample annealed 750°C for 48 h. (After ref. 10).

The most recent report from Amherst[10] returns to the $Nb_3(Al,Si,B)$ material mentioned earlier. In this latest work, the crystallization process, the key both to ductility and to high critical currents, was studied in more detail. It turns out that the crystallization temperature is crucial. If it is too low (below 750°C) heterogeneous nucleation intervenes near the ribbon surface and the grain size is non-uniform; also admixtures of unwanted crystal structures were found. The ribbon thickness, which determines the quenching rate during vitrification, also affects crystallization on subsequent reheating, and the thinnest ribbons (~ 5 μm thick) gave the most uniform A15 grain sizes. The thinnest ribbons, heated at 750-780°C, crystallized via homogeneous nucleation to a uniform grain size of 30 nm, consisting of almost 100% A15. The critical currents measured with this tape are shown in Figure 2.

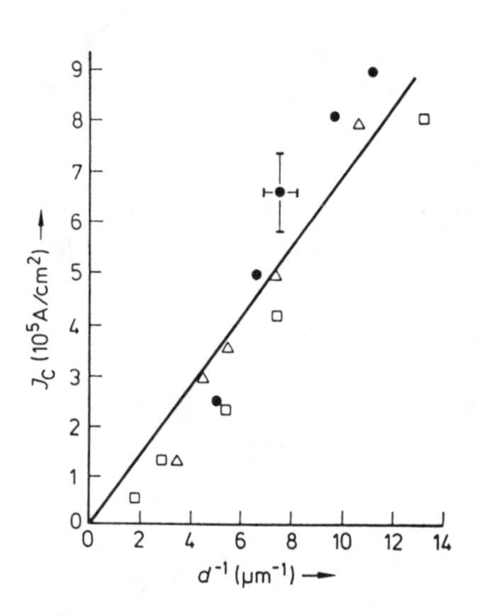

Fig. 3. Critical current densities at 4.2 K and a field of 4 T for various A15 compounds as a function of inverse grain size. Bars indicate estimated errors. • Nb_3Sn. □ V_3Ga. △ V_3Si. (After ref. 11).

Several years ago, Livingston[11] showed that the grain size, acting through flux-pinning at grain boundaries, is the chief determinant of the critical current in the A15 compounds, which cannot be cold-worked to introduce additional flux-pinning centers. He showed that the relationship between reciprocal grain size and critical current at a fixed field for different A15 phases falls on a master line (Figure 3). If one extrapolates this line to a 30 nm grain size, the predicted critical current, $\approx 3 \times 10^{10}$ A m^{-2} at 4 T, is less than half of the value achieved at Amherst. It would be interesting to examine the powerful performance-enhancing effect of ultrafine grains in terms of theories of pinning force densities[12] and also in terms of theories of scaling laws for flux-pinning[13]... but that is for specialists. Meanwhile, rapid solidification of a series of complex alloys has demonstrated, as so often before, that by this approach, the unexpected should be expected.

1. G. Sauthoff, *Adv. Mater.* **1**, 53 (1989); *Angew. Chem. Int. Ed. Eng. Adv. Mater.* **28**, 243 (1989).
2. H. Warlimont, *Adv. Mater.* **1**, 225 (1989); *Angew. Chem. Int. Ed. Engl. Adv. Mater.* **28**, 947 (1989).
3. C.C. Koch, *Int. Mater. Reviews* **29**, 201 (1989).
4. K.H.J. Buschow, *Mater. Sci. Repts.* **1**, 1 (1986).
5. L.R. Gresock and M.T. Clapp, *Mater. Lett.* **2**, 492 (1984).
6. M.T. Clapp and D. Shi, *J. Appl. Phys.* **57**, 4672 (1985).
7. M.T. Clapp and D. Shi, *Appl. Phys. Lett.* **49**, 1305 (1986).
8. T. Manzur and M.T. Clapp, *J. Appl. Phys.* **65**, 2384 (1989).
9. R.W. Cahn, *Materials Research Society Symposium Proc.* **81**, 27 (1987).
10. M.T. Clapp, Zhang Jian and T. Manzur, *J. Mater. Research* **4**, 526 (1989).
11. J.D. Livingston, *phys. stat. sol. (a)* **44**, 295 (1977).
12. J.E. Evetts, A.M. Campbell and D. Dew-Hughes, *J. Phys. C* **1**, 715 (1968).
13. E.J. Kramer, *J. Appl. Physics* **44**, 1360 (1973).

Materials Conferences

71

Sticks and Carrots

Nature, **248**, 640 (1974)

At a conference on the conservation of materials, sponsored by the Institution of Chemical Engineers and the British National Committee on Materials, on March 26-27 at Harwell, the emphasis was partly on primary metal supplies and the associated topic of substitution, with a bow in the diretion of plastics. Much of the conference however, was targeted on the problems, both technical and political, associated with recycling. The political interest in this field was signalled by the opening addresses delivered by the Chief Scientific Adviser to the Cabinet, Sir Alan Cottrell, and by the Liberal peer, Lord Avebury.

What was for this observer the most interesting issue discussed at the conference was raised by P. F. Chapman (Open University) — a most cogent and persuasive speaker. His primary concern is with the energy content of primary metals implied by the energy input of the extraction and purification procedures. When this aspect is pushed into the forefront, then the systematic recycling of 'energy–intensive' metals becomes particularly attractive. The problem is, how are the public and manufacturers to be persuaded to take the necessary steps — domestic conscientiousness and apposite engineering design—to increase recycling substantially?

Chapman quoted the parable of the common. A piece of common land can support twenty-five cows, and five smallholders each have the right to graze five cows there. All is well until one of them adds a sixth cow, to increase his income a little. All twenty-six cows go a little short of grass, including the rogue smallholder's sixth cow, but overall he gains; most of the loss, without any compensating gain, falls on the other four smallholders. In turn, they add cows, until in the end all are worse off, and the last state is worse than the first for all. This might be a parable for the evils of inflation, but in the present context the point is that the incentive for self-restraint is merely collective, whereas each individual in isolation has a financial motive for selfishness.

A case in point is the design of beer cans. Many of these are made of steel, with a top made of aluminium for its easier tearing qualities. Such cans are almost useless for recycling, since aluminium is an unacceptable contaminant of scrap steel. The manufacturer has an incentive to design a can in such a way that the consumer is most likely to buy it (I take it as axiomatic that most canned beer is of indistinguishable and mediocre quality!). The scrap merchant or the British Steel Corporation gains from a prevalence of aluminium-free cans, whereas the beer drinker and the can maker do not, and may even suffer inconvenience.

This last point, the fact that the financial benefit of improvements in the recycling technique does not go to the people who need to modify their behaviour to bring the benefits about, was also made forcefully by J. B. J. Williamson (Guest, Keen and Nettlefolds Ltd). He words it differently: in the course of materials recycling, ownership of raw material changes from manufacturing industry to the user, until it becomes waste and belongs to everybody and nobody. This is why recycling in the manufacturer's plant ('prompt recycling') is so much more effectual than 'post-consumer recycling'. A recycling industry is defenceless against the vagaries of the public, and is so vulnerable to a fierce price-war with primary producers that it would have to concentrate totally and would have no scope for diversification — hence would be unattractive for investment. This difficulty is enhanced by the fact that technological improvements by users reduce demand for metals in short supply; thus a new, effective method for joining aluminium cable ends reduced demand for copper cables.

Yet an autonomous recycling industry is bound to come. Williamson was persuasive in arguing that the economic forces of the market will not favour recycling; but recycling will become much more attractive if (1) foreign primary sources become less reliable (he had disruption in mind rather than nationalisation), and (2) governments convert environmental priorities into fiscal incentives and threats.

This will not be easy — Chapman suggested that the building of aluminium smelters in Britain with government backing and incentives led to a reduction of aluminium recycling rather than a reduction in primary imports — but it is essential. Plainly, manufacturers and consumers alike must either be cajoled with allowances and other financial incentives (for instance, to use all-steel or all-aluminium cans, and then to collect them after use) or they must be threatened with penalties if they do not behave. The trouble with the stick is that civil servants have to think up a complete programme of legislation and the police have to enforce it. Yet such a programme can only be based on present technology: as Marx commented, the problem is not to understand the world but to change it, and the advantage of using the carrot instead of the stick — incentives rather than threats — is that industry would have a motive for developing new recycling processes and new manufacturing designs congenial to recyclers. There must be a mixture — but let there be incentives for carrot growers.

Sir Alan Cottrell remarked that the British Government will usually act if they have clear, unambiguous information. It is to be hoped that the Civil Service will provide that information and that legislation will follow.

72

Rock Fabrics

Nature, **256**, 459 (1975)

The British Standard Conference is organised by a learned or professional society, lasts 2.5 days, has two parallel sessions, costs £ 25 in registration fees alone, involves a pricey conference dinner, is attended mostly by academics and industrial scientists whose institutions are still willing to pay up, and evokes a British Standard Response to the effect that, with luck, just one paper will be really interesting to the British Standard Conferee. I claim a little artistic licence — the British Standards Institution has not yet, so far as I know, laid down a standard for the BSC — but all seasoned scientists will recognise the gestalt.

Some weeks ago, in mid-May, I had the good fortune to take part in a very different kind of conference in the Geology Department at Imperial College, London. It lasted one day, cost £ 0.50 to attend, had a single session, and was organised entirely by a graduate student, K. McClay, on behalf of a group of graduate students and university staff (the Tectonic Studies Group); the audience consisted of a sprinkling of British and foreign academics and a lot of students, post- and undergraduate alike. The topic was "Fabrics and Textures in Rocks".

The term 'fabric' in a petrographic context denotes both preferred orientations in populations of crystal grains (what metallurgists term 'texture') and the shapes and mutual dispositions of matrix grains and subsidiary phases (what metallurgists term 'microstructure'). Texture arises from mechanical deformation of rocks under pressure (deformation textures) and resultant recrystallization when the temperature is high enough (annealing textures). The processes involved must be related to those that generate deformation textures and annealing textures in metals; the Tectonics Group evidently recognise this, because the two opening speakers (I. L. Dillamore of the British Steel Corporation and R. W. Cahn of Sussex University) were invited to outline the present state of metallurgical knowledge on the two kinds of textures. The geologists present showed an impressive familiarity with the mathematical techniques used to interpret the genesis of deformation textures and the processes that arise during annealing of deformed metals. Thus G. Lister of the Geological Institute, Leiden, dealt with deformation textures in quartzite and S. White of Imperial College with recrystallization mechanisms in the same rock, both drawing extensive metallurgical parallels.

Other contributors (E. Rutter and W. Shaw of Imperial College and R. G. C. Bathurst of Liverpool University) reviewed textures and fabrics of limestones, dolomites and marbles. D. J. Barber and H. R. Wenk (Essex University) applied electron microscopy to these rocks; this technique has only recently proved feasible for petrography, following the introduction of the ion-beam thinning technique, and it seems likely to prove as fruitful here as it has done in metallurgy.

The petrographer has the advantage over the metallurgist that he can examine his (optically anisotropic) minerals by transmitted polarized light, with the aid of a universal stage. Students of quartzite use this approach to determine a 'partial texture', that is, the distribution of the orientations of *c* axes both from one grain to another, and between different parts of the same grain: in this way, detailed textural and micro-structural information on a population of over 100 grains can be obtained within a day's work — information that would take a metallurgist, working with X rays and reflected-light optical microscopy, or with electron microscopy, many weeks to assemble. Dr White in particular, using the *c* -axis technique, was able to draw on metallurgical experience to prove the close similarity of the processes that generate high-temperature textures in quartzite.

The remaining speakers, from Leeds and Leiden, dealt with more complex minerals, such as slates and peridotites. Here again, metallurgical processes such as strain-induced grain-boundary migration were identified.

One of the quartzite specialists spoke a few weeks later at a British Standard Conference (good of its type!), devoted to textures in metals: to a witness who was able to compare the two occasions, it was plain that the geologists were much more ready to learn from metallurgical insights than metallurgists were willing to interest themselves in geologists' problems and techniques. This is a pity, because the two groups have much to contribute to each other.

73

Materials Science and Engineering

Nature, **266**, 777 (1977)

"A general survey or reconsideration" is one of the meanings attached by the Shorter Oxford English Dictionary to the word *Review*. That usage dates from 1604, when modern science was in its birth-throes. The natural philosophers of that time knew and corresponded with each other, and that curse and blessing of our day, the scientific journal, was as yet unknown and unnecessary. Today, the general survey or reconsideration is as essential to the working scientist as the Good Food Guide to the serious gourmet.

A review, of course, tells a scientist in a nutshell what the current state of knowledge is in a field, with extra emphasis on recent facts and insights. Who ever heard of a survey of ignorance? Yet that is what the editor- in-chief of the journal *Materials Science and Engineering*, together with his Advisory Board, set out to provide to mark the tenth anniversary of the journal's publication. The 264 pages of volume 25, dated September 1976, are made up of 35 short essays devoted to "what we do not know about" a range of subjects which fall within the scope of the journal. The tacit objective seems to be to guide the research worker who is contemplating a change of field, or of emphasis within a field, into fruitful problems. The editor invited a number of specialists all over the world to map the gaps and chasms in their knowledge. The result is unconventional, variable and useful.

One contributor complains: "Throughout my professional life I tried to teach my students, when they were writing their first paper, not to elaborate too much what they didn't know. Of that, I told them, the future readers know enough themselves." However, the wise choice of expert contributors renders this review of ignorance acceptable and indeed useful. It is necessary to know a great deal before the gaps in that knowledge can be a source of wisdom. Nobody attempted to restrict himself to a survey of gaps: all, with greater or lesser terseness, first surveyed what is known. As another contributor expressed it —

"I take it that we are asking: What principles governing ... are generally agreed by the scientific community? What principles are controversial? What areas are unknown to the extent that no principles have been formulated?" This contributor achieved something very rare and indeed paradoxical in a scientific review: he included not a single reference to published work, but filled his own prescription in a discursive way, emphasising the important points without documentation. The essay — L. M. Brown's on the Deformation of 2-Phase Alloys — is a thoroughgoing success. This approach can be followed successfully only with a widely familiar topic; one which is less familiar may need plenty of documentation, and yet effectively pinpoint the crucial areas of ignorance; D. A. Vermilyea's essay on Corrosion is a case in point.

Some reviews are highly theoretical and can only be followed by a reader who is already highly expert (for instance, N. E. Cook on Continuous Transformations; M. Wuttig and T. Suzuki on the Martensite Transformation); others are severely practical, addressed to specific industrial objectives (A. G. Chynoweth on Fiber Lightguides for Optical Communications; M. C. Flemings and others on Rheocasting; J. A. Manson on Modification of Concretes with Polymers). Most cover their subjects in breadth, some (N. Brown on The Effects of Gaseous Environments on Polymers; Flemings on Rheocasting) describe only their own work and the problems uncovered thereby, justifiably since they opened up new fields. One particularly illuminating survey, V. F. Zackay's on Thermomechanical Processing, is not so much concerned with gaps in scientific understanding as with the problems of finding a development strategy which will increase the industrial acceptability of the complex form of processing with which it deals. Altogether, the volume is an intriguing study in styles. In general, the broader the author's range of vision, the clearer his presentation. One might have expected that when subject to such severe constraints on the number of words, the narrow specialist would have fared better, but indeed the general principle remains true that one sees the wood better from a helicopter than from the jungle trail.

The Editor does disclaim any attempt to cover the whole spectrum of materials science and engineering, but it is nevertheless true that coverage is lopsided. Nearly 80% of the reviews deal with metals and alloys, and within that category, there is heavy emphasis on the admittedly crucial topic of phase transformations. It is disappointing that only two essays deal with polymers (and one of these only does so tangentially), one with ceramics, and one with oxide glasses (again tangentially). There is no attempt to shirk topics in which understanding is unsatisfactory (E. Rabinowicz has plenty of ignorance to report on Wear, and does so without obfuscation), but no broad topic such as

Materials in Biomedical Engineering is included.

In spite of the inevitable limitations and the uneven quality of the reviews, this volume deserves to be widely purchased by junior research workers particularly (the price is set low to enable them to do so). They should use it in conjunction with another compendium of useful short reviews, *Annual Reviews of Materials Science* , to gain a first-class overview of the direction in which the discipline is currently heading.

Miscellaneous

74

Structure

This was written in 1972 as a specimen article for a projected *Dictionary of Scientific Concepts*, of which the author was a co-editor. Owing to unforeseen circumstances, the project was later aborted. (The numerous terms in italics were marked out for definition or special emphasis).

Two constituents are essential if a structure is to be created: building blocks and laws of assembly. The metaphor 'building block' is appropriate because *structure* stems from the Latin *struere* , to build. The concept applies equally to natural entities, whether inanimate or living, and to artefacts; many natural entities and artefacts can only be understood, beyond a very superficial level, in terms of their structures. Examples of which this is especially true include: an atom, a crystal, an alloy, an organic molecule, a skeleton, an enzyme, a leaf, a nerve cell, a Gothic cathedral, a crane, a society. To say that one understands these things in terms of their structure implies the belief that their properties — physical, chemical, mechanical — or else their functions, in a biological, sociological, or engineering sense, can be explained more or less completely in terms of the characteristics of their building blocks and the way they are put together.

Crystal structure will serve as an example with wide ramifications in several sciences. Any solid element, pure chemical compound or solid solution ideally consists of a *crystal* or an assembly of crystals which are internally identical. The qualification "ideally" is necessary because some liquids are kinetically unable to crystallise and turn instead into *glasses*s, which are congealed liquids. Any crystal has a structure, which is defined when (1) the size and shape of the repeating unit, or unit cell, and (2) the positions and species of each of the atoms located in the unit cell, have all been fully specified. In a perfect crystal, all these features are identical in all unit cells, and the cells themselves are stacked together with perfect regularity in a pattern which is determined by their shape. At one level, the unit cells are the building blocks; at another level, the individual atoms are. The laws of assembly are implicit in the laws of interatomic force: once we know how the strength of attraction or repulsion between two atoms depends on their separation; what angles are stable between covalent bonds issuing from the same atom; how the strength of the force between two complete molecules depends ontheir relative position, then in principle we can predict the entire arrangement of the atoms in the unit cell. Once the crystal structure is established, many physical properties such as cleavage, melting-temperature, thermal expansion, are thereby determinate. In the fullest sense, structure determines behaviour.

In historical fact, many thousands of crystal structures have not been predicted but were empirically determined by x-ray diffraction. The structures fall into distinct families, and the regularities linking and distinguishing different structures form the subject-matter of the science of *crystal chemistry* , which interprets crystal structures in terms of atomic sizes, electrical charges and the nature and strengths of interatomic bonds.

The taxonomy of large families of compounds such as the crystalline silicates, which at one time defied rational interpretation, becomes clear in the light of their crystal chemistry: here the unitary building block is neither the single atom nor the complete unit cell, but an intermediate unit, the SiO_4 tetrahedron — a group of electrically charged atoms (ions) consisting of a central silicon atom surrounded by four oxygen atoms at the corners of a regular tetrahedron. The tetrahedra can share corners, sides or faces, or there can be a combination of these modes of sharing, to form systematically the wide variety of silicates, following well understood rules of electrical neutrality and packing of atoms of various sizes into available interstices. Once a crystal structure has spontaneously formed, it remains stable unless the crystal is heated to its melting point or to a polymorphic transition point if it has one. However, the changes of structure which a crystal can undergo at transition points (that is, its *polymorphism*) are of the greatest scientific interest and practical significance.

No crystal is entirely perfect in its periodicity: in particular, all crystals have a very small proportion (typically one in 100 000) of their atoms missing, forming *vacancies*. The existence of vacancies in concentrations depending on temperature is thermodynamically necessary. Apart from such *point defects* , there are also *line defects* , called *dislocations* . It is a characteristic feature of crystals that some of their properties depend very sensitively on these very small concentrations of defects: Zwicky distinguished *structure-sensitivve* properties such as diffusion, mechanical behaviour and electrical conductivity, from *structure-insensitive* properties such as thermal expansion and refraction of light. For instance, mass-transport, or diffusion, of most kinds of atom through a crystal depends entirely on the presence of vacancies, which alone allow atoms to migrate from one stable site to a neighbouring site. This is because each atom is imprisoned in its site by the close press of all its neighbours. It is only when one of its neighbours is missing that it can jump into the empty site; thereby the empty site has moved to the place where the original atom was, and another atom can now move into the empty site in turn. In this way, a few empty sites (vacancies) can have a disproportionately large effect. The vacancy, which merely denotes the absence of an entity (an atom) from its proper place, is itself an entity with clearly defined properties. (There are numerous examples in science of such non-entities, which are nevertheless entities in their own right — for instance, metaphorical 'holes' in electronic energy bands where a particular energy level in a quantised series of energy levels is missing).

Many common crystals, such as for instance metals and salts, are in effect giant molecules. One cannot pick out any one pair of sodium and chlorine ions in a rocksalt crystal and pinpoint that pair as a distinct molecule. The long-established paradigm of the individual molecule was rudely disturbed by the determination of early crystal structures such as sodium chloride. This salt is a giant molecule in the crystal form and dissociates into its components in watery solution: the NaCl molecule never exists at all, and so it is meaningless to inquire into its structure. However, other crystals, especially organic ones (for instance, tartaric acid) *are* built up of individual identifiable molecules, which retain their identity even when the crystal dissolves or melts. The geometrical arrangement of the atoms in such a molecule is a central study of modern chemistry: it is the subject-matter of the science of *structural chemistry*. Many organic molecules can adopt different configurations without change of composition — different *isomers* — and such isomers may have quite different properties. Here again, structure determines behaviour.

Once X-ray diffraction was discovered in 1912 and crystal structures could be determined, it became clear that the external form of an individual crystal is determined primarily by its (atomic) crystal structure. The external mirror-smooth faces of a crystal echo the internal planes of atoms. Nevertheless, a closer examination shows that the circumstances of crystallization also affect the visible form of a crystal: the speed and direction of growth of a crystal growing from solution can variously make it long and spindly, tablet-shaped large, or microscopic. The familiar dichotomy between genetic and environmental influences on the characteristics of organisms is equally applicable to inanimate crystals.

Structure in a somewhat different sense is a central concept in the science of metallurgy and its modern offshoot, materials science, and in the related science of mineralogy. One is here concerned with *microstructure*.. Alloys, ceramics, minerals (in short, most materials) consist of small crystal *grains* , either of a single crystalline species or else of two or more distinct species, each constituting a *phase*. The compositions, sizes, shapes and mutual geometrical disposition in space of these grains are the variables of a microstructure, and in their totality determine the properties of the material. To take a simple example, a pure metal consists of grains, typically ≈ 0.1 mm across, abutting on each other at plane or slightly curved grain boundaries. This simple kind of microstructure results, for instance, directly from the casting process. Unlike a crystal structure, however, this kind of *polycrystal* is not stable, because the three boundaries meeting at an edge, or the four meeting at a point, cannot be in equilibrium at all such edges or points (they can be called nodes, collectively). At most, such equilibrium will obtain at a minority of nodes. An example of a stable node would be an edge with three identical boundaries meeting along a straight edge at mutual inclinations of 120°; since the interfacial tensions are equal, they will be in balance for this very special configuration only.

If a polycrystal is heated to a temperature high enough for atomic diffusion to proceed rapidly, then atoms at grain boundaries can change allegiance from one grain to its neighbour and the grain boundaries migrate, in an attempt to achieve equilibrium at the various nodes. The situation is exactly analogous to a liquid foam, as for instance the head of a glass of beer, The foam steadily coarsens but never reaches a stable condition. The polycrystal or the foam resembles a human society: in repairing a disequilibrium in one place, it creates one at a nearby point, and universal harmony can never be attained. For this reason, some grains are extinguished, others grow, but equilibrium is never attained and the process continues indefinitely.

The topology of single-phase grain structures (closely related to cell assemblies in living plant and animal tissues) has been extensively studied in terms of the statistical relations between the numbers of grain interfaces, edges and vertices, which in turn depend on the grain-size distribution and on the degree of elongation or flattening of grains. For instance, statistical examination of grains of beta-brass, of certain plant cells, and of bubbles in a foam have shown that in all cases, the most probable number of edges per cell or grain face is about 5, with a statistical variation in the range 3 – 8. Likewise the most probable number of faces per polyhedron is about 13.5, and the most probable number of corners per polyhedron is about 23. All these topological relationships can in principle be interpreted on the basis of the balance of forces at different types of nodes; the distinguished microstructure expert, C.S.Smith, was the first to analyse these relationships rigorously, and showed how the frequency distribution of edges per grain, for instance, can be determined from first principles. The study of such relationships in both materials science and biology is at present a lively pursuit, and forms part of the new science of *stereology* (sometimes called *quantitative metallography*).

When at least two distinct phases of different compositions exist in a microstructure, the range of possible microstructural types increases sharply. Types include: a eutectic structure, consisting mostly commonly of parallel arrays of successive lamellae of the two phases; a Widmanstätten structure, consisting of platelets or needles of one phase aligned along particular crystal planes of the other phase, termed the matrix; a globular distribution, consisting of spheres of one phase (with a range of sizes) arranged at random in a matrix; a spherulitic structure, each spherulite consisting of laths radiating from a common centre. Each of these is subject to microstructural variables such as the absolute scale of the structure, relative thicknesses of the two kinds of lamellae, mean free path between spheres, and so on. These variables in turn are determined by physical properties of the constituent phases, of which the most all-pervasive is the *interfacial energy* between grains of two distinct phases, or of the same phase. Interpretation of microstructures is increasingly resolving itself into theories based on the relative magnitudes of the interfacial energies involved in a microstructure.

Several of the structures referred to, such as spherulites and lamellae, are found in *polymeric* systems (plastics in popular parlance). The details of microstructure in these systems are however determined by quite different considerations, because the building blocks are not individual crystal grains but rather the long, flexible polymer chain-molecules which interact with each other in special ways. Polymeric systems form a structural link between living and inanimate systems.

Microstructures are formed not only by solidification from the melt or crystallization from liquid solutions, but also by *phase transformations* in the solid state, of which there is a great variety of types. For instance, one of the principal mechanisms for hardening an alloy is to arrange to deposit sub-microscopic particles of one phase inside grains of another phase; this is typically done by heating at a temperature at which the solubility of the second phase is exceeded. — The detailed study of solid state transformations and the heat-treatments required to bring them about is a central concern of metallurgy and materials science, and also enters the study of mineralogy (here subterranean pressure is a further variable).

Crystal structures and microstructures are both involved in the complete description of a material: note, however, that they operate at different levels and in fact exemplify a *hierarchy of structures*. Crystal structure and molecular structure involves architecture at the atomic level, whereas the units of microstructure — grains — embody typically about 10^{15} crystalline unit cells each. The engineer then assembles pieces of material into an engineering structure such as a bridge or crane. Each hierarchy has its own regularities and also its own types of defects, and the properties associated with both regularities and defects at one level govern the laws of assembly and the properties of the next level up. The similarity of such a hierarchy to what is found in a biological system — proteins, cells,organs, organism, animal and plant populations — is evident. Just as recognition of the universality of biological hierarchies has led to an erosion of the boundaries between the classical biological fields of study and their partial replacement by general fields such as molecular biology, histology and ecology, so the study of crystal structure and microstructure (and their modes of genesis) plays an increasingly central part in the education of metallurgists, materials scientists and mineralogists. Indeed, the concept of a hierarchy of structures is of very wide application, and spills over into the social sciences. Thus, society is organised in interlocking hierarchies of individuals, families, professions, work-groups, language-groups, nationalities, and the behaviour of one of these collective entities is determined in part by the other levels in the hierarchy.

Crystal structures at one extreme and engineering structures at the other differ in one vital respect. Crystal structures are determinate once the constituents and treatment have been specified, whereas engineering structures are designed by human intelligence to serve a function. Microstructures are intermediate: materials are 'designed' to fulfil a need, but the designing involves a more or less conscious manipulation of the microstructure which determines properties. When these distinctions are transferred to the biological sphere, the temptation to describe structures in teleological terms is ever present. An anatomical textbook will describe the structures of organs and cell constituents in terms of their biological functions: yet one is not tempted to describe the intricate construction of a mica crystal in terms of its useful ability to cleave into thin sheets.

The exceedingly subtle problem of the origins, repetition and assembly of biological structure has been memorably discussed by Monod. He is concerned with proteins, which are polymeric long-chain molecules made up, in enormous variety, from some twenty 'building blocks' or monomers. Proteins are the first steps in the epigenetic sequence by which immeasurably complex structures grow from single constituents. In terms of concepts of atomic fit between various building blocks, Monod is able to show how monomers pick each other out of a copious mixture and assemble in a pre-ordained pattern. The structure of giant protein molecules, as of simple crystals, is implicit in their building-blocks. To quote Monod, "the epigenetic building of a structure is not a *creation*; it is a *revelation*." He goes on to voice the conviction of present-day biologists that the same essential principle applies to the assembly of cells to form tissues, the conviction that different kinds of cells recognise each other and associate spontaneously into predeterminate structures in consequence. The anthropomorphic metaphor of the mutual recognition of building blocks is indeed central to the formation of all spontaneous structures, as distinct from those designed by human intelligence.

If a metaphor is useful in understanding the origin of structures, the concept of structure itself has metaphorical overtones. David Edge points out that 'structure' carries implications of stability and immutability: a building may survive longer than a man or his institutions, and in another sense the structure and therefore the form of a crystal is for ever unchangeable. It is perhaps for these reasons that radicals use 'structure' as a favourite metaphor for persistent institutions and patterns that must at all costs be toppled. The irony of this usage is that structures are in fact among the most labile and changeable of things: not only do organisms grow and decay, but one crystal structure transforms into another as it is heated, microstructures coarsen and alter out of recognition. What is stable is not the actual, but the potential.

BIBLIOGRAPHY

R. C. Evans, *An Introduction to Crystal Chemistry* (Cambridge University Press, second edition, 1964).
A. F.Wells, *The Third Dimension in Chemistry* (Oxford, Clarendon Press, 1955).
C. S. Smith, Some Elementary Principles of Polycrystalline Microstructure, *Metallurgical Reviews*, **9** , 1 (1964).
E. E. Underwood, *Quantitative Stereology* (Addison-Wesley, 1970).
D'Arcy W. Thompson, *Growth and Form* (Cambridge University Press, 1942, 1961).
L. L.Whyte, *Accent on Form* (Harper & Bros., New York, 1954).
L.L.Whyte, A. and D. Wilson, *Hierarchical Structure s* (New York, 1989).
J. Monod, Chapter 5, on Molecular Ontogenesis, in *Chance and Necessity* (Collins, 1972).
D. O. Edge and J. N.Wolfe (editors.) Chapter on Technological Metaphor, in *Meaning and Control* (Tavistock, 1973).

75

Figures of Merit

For details, see Article 74. (1972).

An engineering design is always a compromise. In a motor-car, acceleration has to be balanced against petrol economy; in an electric light bulb, high efficiency against long life; in a communications link, reliability against cost. Generally, to arrive at a compromise involves a combination of calculation and testing, and the calculations are themselves often involved and inexact. There are, however, situations in which an optimum compromise can be derived from a simple algebraic formula: this is a *Figure of Merit.*

A good illustration of this is the design of an electric capacitor (sometimes called a condenser). Assume that the test of fitness is that the capacitor should be capable of storing a given amount of energy at a given voltage V, while occupying the smallest possible volume. The stored energy = $\frac{1}{2}CV^2$, where C is the capacitance. $C = \varepsilon A/t$, where ε is the dielectric constant of the dielectric substance between the plates of the capacitor, A is the total area of the dielectric and t its thickness. t cannot be indefinitely reduced, since the dielectric has a breakdown field $E_b = (V/t)_{crit}$, at which the capacitor will catastrophically short-circuit. Assume that t is reduced to this minimum permissible value, t_{crit}.

The volume of the capacitor, $v = At_{crit}$. Eliminating A and t from the various relations quoted, one obtains

$$v = \left(\frac{Ct_{crit}}{\varepsilon}\right) t_{crit} = \frac{C}{\varepsilon}\left(\frac{V}{E_b}\right)^2 = CV^2 . \frac{1}{\varepsilon E_b^2}$$

The stored energy, $\frac{1}{2}CV^2$, is specified, so that the criterion of fitness (smallest possible v) becomes the largest possible value of εE_b^2. This product is the figure of merit. Both ε and E_b should be as high as possible, but since E_b is squared, evidently a marginal improvement in breakdown voltage, E_b, is preferable to a comparable improvement in dielectric constant, ε. The important point is that a single algebraic expression identifies both the crucial variables and their functional relationship.

A necessary (but not sufficient) condition that a figure of merit should exist for a particular problem is that the criterion of excellence should be clearly defined. Here we have taken the criterion to be the smallest possible size. This can be justified in electrical engineering terms because a small capacitor can discharge faster, has a smaller stray inductance, etc. Apart from that, a small capacitor evidently requires less dielectric and thus is cheaper (provided manufacturing costs do not increase with progressive miniaturisation). If the capacitor is needed for airborne equipment, then a reduction in weight has a quantifiable cash value. The tacit hypothesis behind the figure of merit, εE_b^2, (which has to be maximised) is that the smallest capacitor is best for a given ability to store energy.

Another important figure of merit in the field of electronics is that for a thermoelectric generator material. A thermocouple is the simplest example of such a device, in which two dissimilar materials are welded together in pairs so that one set of junctions is heated and the other set cooled, and electric power at voltage V is generated. A high thermoelectric coefficient, α (=V/ΔT), where ΔT is the temperature difference between the junctions, is evidently beneficial, and so is a high electrical conductivity, σ, which maximises the current for given values of V and ΔT and a fixed geometry. However, a high thermal conductivity, K, is a drawback, because then the junctions have to be moved further apart to limit heat flow between them and this in turn raises the electrical resistance of the device. A simple calculation leads to $(\alpha^2\sigma)/K$ as the figure of merit; this has to be averaged for the two constituent materials of the device. This way of formulating the performance of a thermoelectric device allowed physicists, at an early stage of research into improved materials, to apply theory to predict which *category* of material would be most promising for investigation, and this led to the examination of semi-metals and semiconductors in preference to metals.

One of the most famous figures of merit in the whole field of engineering is Shannon's formula for the number, n, of independent messages which can be sent in time t over a radio link of bandwidth Δf, in terms of the ratio of signal power, P, to noise power, N. Here "bandwidth" signifies the range of frequencies of the radio carrier wave which is sent out over the transmitter. Shannon's formula is

$$n = (1 + [P / N])^{\Delta f . t}$$

This is the central formula of the Information Theory originated by

Shannon and in its day saved engineers much unnecessary empirical work in the process of improving the quality of radio communications.

Mechanical engineers also make extensive use of figures of merit. For instance, the ability of a flywheel of optimum geometrical design and fixed mass to store rotational energy is proportional to T/σ, where T is the tensile strength and σ the density of the constructional material. Thus a light, strong material proves to be best. The criterion implicit in this figure of merit is that the flywheel should be accelerated until it is just about to fly apart under centrifugal stresses. The material can be selected on the basis of this simple figure of merit, but the optimum geometrical design has to be worked out by successive approximations: there is no simple analytical criterion for this.

The most universally applied of all figures of merit in engineering is Carnot's expression for the maximum fraction, F, of the available heat energy that can be converted into mechanical energy in an ideal engine in which heat enters at temperature T_1 and leaves at temperature T_2:

$$F = (T_1 - T_2)/T_1$$

This perfectly general expression allows the engineer to test the quality of an engine in terms of its approach to this theoretical limiting efficiency. Like all good figures of merit, it indicates clearly in what sense the situation must be altered to improve the outcome: the temperature at which heat is injected into the engine must be as high as possible. Before Carnot's theory, this was not recognised.

A concept closely related to a figure of merit arises in the common engineering practice of modelling. New designs for aircraft are tested by subjecting scale models to a current of air and measuring forces on the model. Ship models are pulled across tanks in which precisely regulated waves are excited. The problem common to all such tests is: How are measurements made on a model scaled up to the real thing? It is not sufficient to argue that a 1:10 scale model aircraft should be tested in a wind blowing at 1/10 the velocity of the true air speed of the full-scale aircraft which the tester wishes to simulate. An expression can be formulated which has to be made numerically the same for the prototype and the scale model if the model is to stand in effectively for the prototype. Thus, the flow pattern of air past a wing is determined by viscous forces, inertia and pressure variations. If these variables are the same at points similarly related, geometrically, to the full-scale wing and its model, then the flow pattern at the two points will be identical. It can be shown that this condition is satisfied if the dimensionless quantity $\rho VL/\eta$ has the same value at both points. Here ρ and η are air density and viscosity, respectively, L is a characteristic dimension (such as the wing chord) and V is the air velocity at the point in question. The expression is the *Reynold's Number*. If ρ and η (dependent on air temperature), and the wind velocity V are suitably adjusted so that Reynold's number is the same for model and prototype, then the measured forces acting on the model can be proportionately scaled up to predict the forces that would act on the prototype. 'Dynamic similarity' is then achieved. It is interesting to note that (for fixed density and viscosity of the air), the air velocity must be *increased* as a model is progressively scaled down.

For ships, water flow patterns are determined by inertia, pressure and gravitational forces. Here dynamic similarity (with respect, for instance, to drag on the ship's hull and response to waves) is achieved if *Froude's Number*, $V \sqrt{(Lg)}$ is equalised for model and prototype. (Here g is the acceleration due to gravity and L is a linear dimension). The Reynold's or Froude's Numbers can properly be regarded as figures of merit for the matching of a model to a prototype.

It is possible to apply the concept of a figure of merit to biology, as a somewhat speculative aid to interpreting evolutionary and physiological developments. John Maynard Smith, for instance, has calculated the variation of the power output needed by a bird for hovering flight as a function of its size. For a number of good physiological reasons, he argues that maximum power output increases only as L^2 (when L is a linear dimension of the bird), and not as the volume (proportional to L^3). Applying simple aerodynamics to the situation of hovering flight, he finds that the power *needed* for hovering increases as $L^{3.5}$. This is much faster than the rate of increase of power *available*, and so a limiting size is inevitable; this is in fact about 35 lb. This formula relating power needed to linear dimension is a scaling factor and not strictly a figure of merit. If regarded as the latter, it indicates at first sight that flight should be the easier, the smaller the bird. In terms of power requirements alone, this is true, but the problem is complicated if one takes into account capacity for gliding flight, the power needed to sustain basic metabolism, and indeed the frequency with which the bird must find food, considerations which appear to favour larger birds. (The fact that birds have not evolved to a preferred size suggests that, even if we knew how to calculate it, a figure of merit for bird size probably does not exist.) The limitations of this example shows that a figure of merit or related criterion can only be as good as the scientific arguments that are used in its derivation. In engineering as well as biology, figures of merit must not be pushed beyond their proper context.

Maynard Smith also exemplifies a true biological figure of merit – that is, the problem of the optimal gait of a mammal. When travelling fast, a mammal spends part of each stride with all four legs out of contact with the ground. It is possible to analyse, on simplifying assumptions, the power output required for different values of the ratio, j, of time spent 'floating' to time spent in contact with the ground. The result is: $(1 + j) \propto V^2/L$, where V is velocity and L is a linear dimension. At top speed, j is zero for an elephant, 0.3 for a horse and 1 for a greyhound. In fact, j varies less rapidly with size than this formula predicts. In effect, the formula is a figure of merit for greatest energy economy, and observation indicates that optimum gait is influenced by other, unrecognised factors also. Again, a figure of merit is limited by the accuracy of the hypotheses used in devising it.

BIBLIOGRAPHY

C.E. Shannon and W. Weaver, *The Mathematical Theory of Communi cations* (Chicago, 1949).

J. Maynard Smith, *Mathematical Ideas and Biology* (Cambridge, 1968).

76

Boom and Doom Loom

Nature, **239**, 11 (1972)

A recent public venture by Harvey Brooks, professor of applied physics at Harvard University, was a lecture on materials in a steady state world, delivered last autumn to the two leading professional bodies in American metallurgy, and now published in their joint journal[1]. Brooks is a man of unusually wide experience in the borderlands of science, engineering and politics in the United States, and has participated in many enquiries on various aspects of science policy.

He began with a thumbnail sketch of Forrester's world models (the lecture antedated publication of *The Limits to Growth*, which took that model a stage further), and contrasted the underlying outlook of that form of global future-gazing with the changes during the past century in the supply and real price of raw materials, bearing in mind the fact that dire prophecies of the exhaustion of natural resources have been current for a long time; they were particularly prevalent in the first two decades of this century. Among a number of striking observations, he showed that the consumption of materials *per capita* has grown much less rapidly than the gross national product (GNP) *per capita*, and that the output for each man employed has risen particularly fast in the mineral extraction industries. He made the intriguing suggestion that one symptom of the lack of pressure on these industries has been the sharp decline of mining engineering and extractive metallurgy in universities.

Having very fairly presented the optimist's case, Brooks then turned to the other side of the medal. Some materials pose a much more serious problem than others: thus *per capita* use of paper is closely correlated with *per capita* GNP, but grows five times for each doubling of the *per capita* GNP. Moreover, the price in real terms of forest products has trebled over the past century, whereas on the same basis the prices of minerals are unchanged. Brooks went on to classify metal ores in a particularly illuminating way. Some metals, especially iron and aluminium, are present in enormous quantities in ores whose grade varies from those at present exploited, continuously down to very much lower levels. Others, such as copper, nickel, vanadium and lead, occur in ores with a continuous range of grades, including substantial supplies in ores at present somewhat too lean to use, and these will become steadily dearer without, presumably, a state of crisis being reached. The real problem comes with what he calls the "metal vitamins", used as essential alloying elements; these include, among others, mercury, tungsten, tantalum, silver, tin, manganese and molybdenum; some of these of course have 'nutrient' as well as 'vitamin" uses. They are won from rich mineral veins which are fast being exhausted, but once this has happened, there is apparently no supply of progressively leaner ores, only the general background concentration in the Earth's crust, which is orders of magnitude lower.

Many metals can be won for a long time by a combination of improved technology and higher real price, but this will not be true for all. When one of the problem metals also poses a special pollution problem, it will be interesting to see how rapidly substitution or abolition of function results. Mercury is a good example, and it may be significant that the price has recently been plummeting: on the other hand, the influences of cartels among producers and consumers are so strong that it is dangerous to read too much into price fluctuations of metals.

Brooks continued by outlining some of the more novel problems of the future, which are all connected with the increasing difficulty of discerning trends far enough in advance to plan the necessary industrial changes in the extractive industries. He instances sudden unpredictable political changes, including ever stronger cartels of suppliers, and changes imposed by rapidly veering public opinion, which do not allow the necessary lead time for technical change. He also assessed in general terms the problems arising outof the universal drive towards growth,and said quite bleakly that "there appears to be a major discontinuity between current trends and national objectives, in virtually all national economies, and the need to obtain some sort of equilibrium with our environ ment. It is not obvious that the various homeostatic mechanisms of the world economy will operate to set in motion automatically the forces which will propel us towards a new equilibrium. The discontinuity between current trends and goals and ultimate necessities is very hard to think about in a logically consistent way. The apocalyptic rhetoric of the ecological ideologies is not very helpful because they do not tell us how to get there from here, or even how to begin".

In a tentative attempt at propounding some solutions, he urged a government–financed expansion of long-range geochemical and geophysical research "aimed at defining and assessing the future availability of metallic ores", research which cannot fairly be left to mining firms. He thinks that, although substitution of more common for less common materials must continue apace, whole technologies must give way to less materials–intensive ones: he gave as an example advanced telecommunications as a substitute for travel. He also emphasized the role of draconian changes in national laws to oblige all companies to do what no one firm could do in isolation without going out of business, and instanced the new American car exhaust emission laws as an example of what can be achieved. This law may well in retrospect mark a watershed, showing many who now induge in fashionable despair what man can do to help himself.

1. H. Brooks, *Metallurgical Transactions* 3, 759 (1972).

77

Metal Bites Metal

Nature, **265**, 11 (1977)

The recent Report of the Court of Inquiry (HMSO, London, 1975) on the disaster at Flixborough, where a caprolactam manufacturing plant exploded with grave loss of life, cited surprising evidence concerning weakening of hot stainless steel by traces of metallic zinc on the steel surface. This had not been recognised before the disaster inquiry and the special research commissioned to help in that inquiry. It appears that minute quantities of zinc, which can reach the steel either as liquid or as vapour, lead to fracture within seconds at modest stresses (5-6 kg. mm^{-2}) in the temperature range 800-900°C. Without the zinc, a stress well over 10 kg.mm^{-2} would be needed in this temperature range to achieve similarly rapid fracture. This is an instance of stress-corrosion, a phenomenon which involves the mutual abetment of stress and corrosion: the corrosive agent accelerates the action of the stress and the stress enhances the corrosive attack. The importance of zinc-embrittlement in the sequence of events that led to the explosion was the central and highly disputatious issue at the Court of Inquiry; mutterings of disagreement with the official findings may still be heard today.

Stress-corrosion by liquid metals takes place far more commonly than was recognised two decades ago: the subject has been comprehensively reviewed by Kamdar[1]. Generally, such attack is not pronounced at or near ambient temperature, and it is thus a form of brittle fracture, enhanced by adsorption of the contaminant at the root of surface cracks. Some trace metals embrittle a variety of solid metals: thus mercury embrittles aluminium, zinc, cadmium, iron, silver, titanium and others, gallium attacks most of these metals (most spectacularly, aluminium), whereas bismuth only embrittles copper, and cadmium only iron and titanium. Mercury can claim championship status as a general trace nuisance in biology and engineering alike!

A paper by Grubb in this issue of *Nature* [2] adds a further example to the growing number of instances of liquid metal embrittlement. It resembles the stainless steel/zinc syndrome in that the embrittlement is specific to high-temperature stressing and does not arise at room temperature, unlike the general run of such cases. Grubb has discovered that zircaloy-2 (a zirconium-base alloy containing 1.5 at % tin and small amounts of iron, chromium, nickel and oxygen, which has long been used to enclose fuel elements, in water-moderated nuclear reactors) is embrittled by traces of metallic cadmium — solid, liquid, or dissolved in another liquid metal — in a narrow temperature range, 300–340°C. The fracture is purely brittle in character under these circumstances and takes place by cleavage on the (hexagonal) basal plane. The report is clearly preliminary and there is no information yet about the time during which stress must be applied to generate fracture: it is a frequent characteristic of stress-corrosion that fracture is delayed.

This finding follows the discovery, ten years ago, that iodine can cause stress-corrosion failure in zircaloys in the range 250–500°C. Wood[3] studied the matter in detail for zircaloy held at 300° C, which is a typical reactor-operating temperature for water reactors. For adequately large stresses and surface concentrations of iodine, fracture ensued, but previous anneals to remove residual internal stresses greatly reduced its incidence. Iodine-embrittlement is of peculiar concern be cause iodine, a volatile element, is a major fission product and is thuscreated inside the fuel element during service. Other fission products, however, would be expected to 'getter', that is, neutralise any free iodine; different metals can be regarded as being in competition for the attentions of free iodine. There was, in 1972, no clear evidence that any cracks in fuel element containers had been caused by this mechanism, and I have seen no later papers to suggest that any such events had been established. Wood's paper pointed the way to a number of precautionary measures.

Cadmium-embrittlement of zircaloy might conceivably be of consequence in a nuclear reactor because cadmium is one possible constituent of neutron-absorbing control rods, and Grubb's discovery may serve to discourage the use of this particular neutron absorber.

More generally, it is now clear that liquid metal embrittlement (if we may regard iodine as an honorary metal for present purposes) falls into two clearly distinct categories: hot or creep embrittlement, and cold embrittlement. It will be the next task of background research in this field to establish the characteristics of aggressor/victim combinations that predispose a system to one of these two modes of attack.

1. M.H. Kamdar, *Progr. Mater. Sci.* **15**, 289 (1973).
2. W.T. Grubb, *Nature* **265**, 36 (1977)
3. C.J. Wood, *J. Nucl. Mater.* **45**, 105 (1972).

78

Transformations

An Encyclopedia of Ignorance (Ronald Duncan and Miranda Weston-Smith, editors) p. 136 (Pergamon Press, Oxford, 1977)

Circe, the enchantress, turned men into swine: but the beasts, as they snuffled among the acorns, wept for their lost human forms, for they preserved the minds of men. The change of outer form while some inner essence is maintained intact — the process of *transformation* — is a recurrent theme in literature, mathematics and science alike. In science, the notion of transformation is linked with that of *structure*, for structure determines appearance. The bones of the skull fix the features, the sequence of amino acids specifies the gene that codes for eye colour, the arrangement of atoms or molecules in a crystal determines its shape. There is, however, a crucial distinction between the gene and the crystal. It is of the essence of a gene that it is almost always invariant and replicates precisely true to type; when it does undergo a minute mutation, that in turn becomes invariant, and the resultant biological change with it. Not so with a crystal: the array of atoms is labile and may change reversibly from one pattern to another.

Thus a crystal represents something special in nature, for it is at the same time a single substance and two or more substances. When iron is heated, a well-defined temperature (910°C) is reached when the stacking of iron atoms all at once changes; in crystallographic language, it changes from body-centred cubic to face-centred cubic. On cooling, the structure changes back, and this change on cooling brings with it changes in properties (Fig. 1). In pure iron these changes are trivial, but dissolve a small amount of carbon in the hot iron and the resultant alloy — a simple form of steel — transforms with a dramatic change of properties. The previously soft and pliable steel becomes very hard and brittle.

Not all crystals behave in this fashion. Some transform, sharply, at one or more well-defined temperatures. Some (including steel) transform over a range of temperatures. Some transform suddenly and others take their time. Some crystals can transform only if the pressure to which they

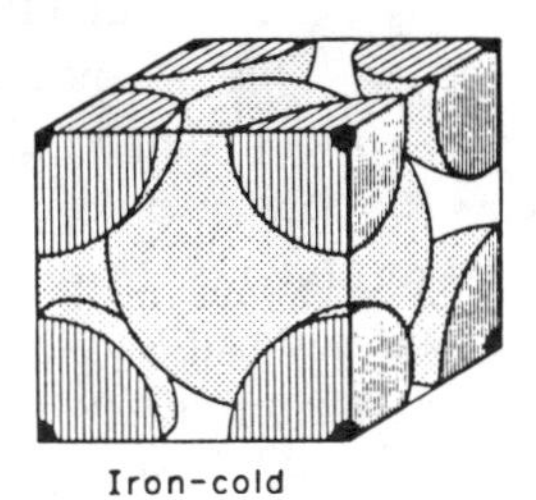

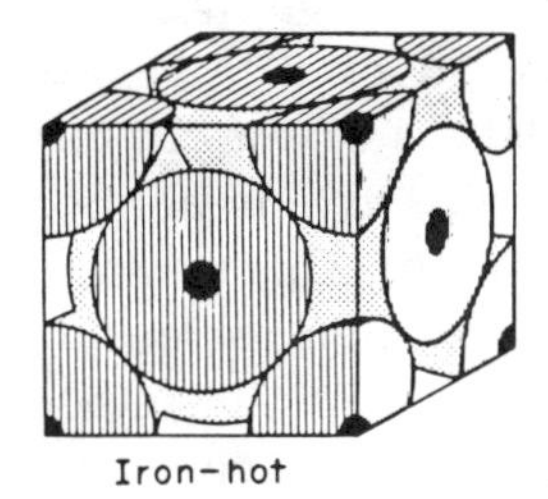

Fig. 1. The crystal structures of iron below and above 910°C. Each sphere represents an identical atom.

---

are subjected changes: the most celebrated instance of this is the conversion of almost pure carbon in the form of graphite into the form of diamond when it is heated under a very large confining pressure. Still others cannot transform at all; melting, that is, the loss of all crystaline order, preempts any change of crystalline pattern. The generic name for transformations in crystals is 'polymorphic transitions'; the Greek word implies the existence of many *forms.* The concept of polymorphism, however, implies that some feature is common to all the forms: there is a unifying essence, just as there is between the grub, the pupa and the moth. What is it?

It is easy to say that the chemical identity of a polymorphic crystal is invariant: but what does the term mean? A chemical substance is defined by the nature and proportions of its constituent atoms and the way they are assembled. Two organic chemicals can be made up of the atoms in the same proportions in each molecule, yet be put together differently and have quite disparate properties. Such *isomers* are not in general mutually convertible. So the whole molecule is not likely to be the invariant feature in a polymorphic crystal and indeed most such crystals do not contain recognisable molecules at all. The archetypal polymorphic crystal is either an element — such as iron, cobalt, phosphorus, sulphur, uranium — or a simple inorganic compound or solid solution, such as $CaCO_3$, ZnS, CuZn, Cu(Al) . Any such substance forms a crystal that is in effect a single giant molecule: one cannot pick out a single zinc and a single sulphur atom in zincblende and call that pair a molecule. The only constant feature in a polymorphic crystal is the heap of atoms, the elementary chemical building blocks. What varies when such a crystal is heated or compressed is the nature of the chemical binding; the strength, length and mutual inclination of the chemical bonds changes and the constituent atoms may cluster and rearrange themselves so that the local composition varies from point to point.

The understanding and control of polymorphic transitions is the central concern of the science of metallurgy. This is inevitable, for metallic artefacts have to be formed — and so must be soft and pliable — and they also have to be strong and hard to withstand the shocks and stresses of service. That paradox can only be resolved by tranforming the structure of the artefact *after* it has been shaped and put together.

The technological importance of transformations in metals and alloys, then, is evident enough and the scientific problems are subtle and varied in the extreme. To appreciate why this is so, it is necessary to invoke another category, that of *microstructure.* Most metallic objects consist not of a single crystal but of an assembly of small irregularly shaped crystal grains, which can only be seen with the aid of a microscope. Generally, more than one species of crystal is present and useful alloys are most often composites of several distinct crystalline *phases.* The sizes, shapes, proportions, compositions and mutual disposition of these phases — collectively, the microstructure — are all variable and subject to control. Heat-treatment, designed to alter the microstructure, is the metallurgist's central skill.

The behaviour of a simple carbon steel — the most important industrial alloy — will serve to exemplify the range of phenomena covered by that deceptively simple term, *transformation.* When the high-temperature form of steel (right-hand sketch, Fig. 1) is slowly cooled, it breaks up on transformation into two crystalline forms: almost pure iron in the form shown on the left, together with a compound of iron and carbon (Fe_3C) with a more elaborate crystal structure. (The high-temperature form of iron can readily dissolve carbon; the low-temperature form cannot dissolve much.) Small crystallites of Fe_3C are independently nucleated in many sites. Part of the microstructure consists of thin plates of iron and of Fe_3C in alternation, and the whole assembly is fairly soft. If the high-temperature form is instead cooled suddenly, the transformation process is entirely different. The dissolved carbon is pinned in its existing sites and cannot segregate, for lack of time, and the alloy transforms by an ordered shift of millions of atoms into a new pattern. In the slow process, atom movements are at first random and uncoordinated: in the fast process, they are disciplined and simultaneous. The terms 'civilian' and 'military' have been applied to the two categories of transformation.

The product of the military transformation, containing as it does a great deal of carbon inenforced solution, is extremely distorted and therefore hard but also incapable of resisting intense shocks. If now this product is tempered by slow progressive heating, a new civilian transformation begins and a succession of iron-carbon compounds is formed in sequence, in the form of minute crystallites. Any desired level of compromise between hardness and shock resistance can be achieved, and different parts of the same object can be made to have quite different microstructures. The classic Japanese sword represents the most sophisticated application of these skills.

Military and civilian transformations each exist in rich variety, with many subtle distinctions of mechanism. Steels in particular form a large metallurgical family because of the variety of alloying elements which can be added to the basic iron/carbon constituents. Some of these elaborate steels are sensitive to imposed distortion. In such steels, transformations are induced when a hot sheet or rod is forced into a changed shape, as happens when it is passed through a pair of rollers. An early form of this was the use of a special steel — Hadfield's manganese steel — for the construction of railway points; every time a wheel crashed against a crossover point, the point became a little harder because of the stress-induced transformation. The study of such thermomechanical treatments is a new chapter in metallurgical research and is at a scientifically most intriguing stage. The strangest variant is the *shape memory effect* . Certain alloys — the alloy NiTi is the best known — can be extensively deformed and then on heating, will return to their pristine shapes. This behaviour is quite different from that of an elastic spring: a spring obstinately returns to its original shape when it is let go, whereas an SME alloy humbly accepts its pummelling and stays put in its new shape. Only when it is heated does it home to its original form, even against a strong mechanical force seeking to prevent it. This mode of behaviour is always based on a stress-induced polymorphic transformation of the military type, followed by a reverse transformation when heat is applied. It is as though the natural pugnacity of an army forced into precipitate retreat could only be regenerated by an exposure to sunshine!

The shape-memory effect is extremely intriguing, both for its engineering applications and because of the difficulty of understanding the long-range forces which powerfully drive the transformed alloy back towards its original shape. A most detailed investigation of the microstructural chariges is necessary in order to come to grips with this phenomenon, which is part of a much larger complex of questions concerned with the interplay of temperature, stressand transformation.

Quite apart from the technological justification, many metallurgists, physicists and chemist have long found this field of study irresistible for its purely scientific attractions. It is a satisfyingly hydra-headed creature: two questions raise their heads for each question that is resolved. For instance, the field of transformations in liquid crystals has arrived in the past few years, as a new branch of physics (stolen while no one was looking from the chemists who opened it up). Liquid crystals are half-crystalline, half-disordered substances that respond sensitively to heat and electric fields: the transformations in liquid crystals from greater to lesser degrees of order have a close family resemblances to ferromagnetic and 'atomic-order' transformations, and are proving amenable to interpretation in terms of a form of 'catastrophe theory' which was applied to ferromagnetism and atomic order long before it was taken up and generalised by mathematicians.

The attractiveness of transformations as a subject for scientific investigation may have something to do with a universal, prescientific human obsession, attested by much ancient legend and folklore. The Greeks told tales of Proteus, a sea-god. Men would seek and seize hold of him as he sat on the rocks, in order to force him to grant a wish. The evasive god would transmute through all the varieties of living appearance, many repellent or terrifying, in the hope of frightening off his captor so that he might escape beneath the waves. Only the brave man who dared to keep his grip on the god till he had run through his entire gamut of forms and returned to his own godlike shape was assured of his wish. Each creature has its proper form and any departure from it is an affront to the natural order of things. A man's sense of identity is indissolubly linked to his own physical body: those who destroy this link, like Circe, like the jealous fairy who turned the prince into a toad, have always been seen as wicked and destructive. To retain the burden of identity while suffering a mutation in appearance is one of man's enduring terrors.

Underlying this is the philosophical dogma that in nature, building-blocks determine structure and structure defines appearance. Man can use a pile of stones to build a cathedral or a bank, but a pile of iron sulphide 'molecules' is not subject to man's will and is bound to form a particular kind of crystal, always, with a particular yellowish colour and cubic external form. Snowflakes grow in different patterns according to the change of temperature and humidity, but they are all variations on a strictly constant theme, set by the hexagonal structure of ice. The structure of a simple crystal is implicit in its building block — atoms or molecules — and the predetermined forces which bind them together. Complex biological molecules, proteins. are half-way between the determinate simple crystal and the building whose form is subject to man's free will. The same basinful of amino acids can form a multitude of different proteins according to the template provided. It is still a matter of atomic fit: Monod showed how one amino acid selects another out of a copious mixture and by this means the acids assemble into a preordained pattern. As Monod puts it: "The epigenetic building of a structure is not a *creation,* it is a *revelation.*" This is as true of an element as it is of a protein.

Polymorphic transformations (and indeed the melting of crystals) thus conflict with a deep philosophical sense of fitness, of match between form and essence. Those crystals which transform easily are unstable, not sure of their own proper nature, full of doubt, like man tormented by his Freudian unconscious. It is ironic that the most protean of elements is also the most unstable and dangerous: plutonium has four polymorphic forms. By gaining an understanding of transformations, man masters his fears, keeps hold of Proteus, gains his desires, fulfils his needs. At this level, the philosophical, scientific and technological desires of man all impel him to the same study. The mastery of transformations reassures man, the magician, that Jekyll and Hyde are under control.

79

The Legacy of Xenophon

Nature, **280**, 11 (1979)

The US National Academy of Sciences gave the name COSMAT to a massive study it sponsored of the profession of materials science and engineering (MSE), published in five thick volumes in 1974 and 1975. It is a remarkable compilation; because of its wide terms of reference, and especially because of its historical and international emphasis, it is of value far beyond the originally intended readership in its target profession. The report is difficult to locate and few people in Britain have read it; to make at least a small extract accessible to a wider readership, a learned periodical, appropriately named*Materials Science and Engineering*, has published a special issue entirely devoted to a revised and updated version of two chapters from one of the volumes[1]. [See also article 86.]

The heart of the extract is a forty-page essay, by Melvin Kranzberg (editor of the journal *Technology and Culture*) and Cyril Smith (the world's leading scholar in the history of metallurgy), on "Materials in History and Society", complete with a bibliography of 85 items, which alone would justify acquisition of the volume. The emphasis is predominantly on the history of the use and understanding of metals. The essay is a masterly examination of the ways developments in the manufacture, purification, variegation and fitness for purpose of materials influenced, not only the tenor of daily life, but matters of great social impact such as the spread of agriculture; the balance of affluence between patricians and plebeians (as "aristocratic" bronze gave way to "democratic" iron); the spread of the printed word and the printed image; the materials testing institutions which were a precondition of safe machinery and all that implies. The later parts of the essay are more concerned with the gradual development — remarkably delayed, and significant only in the twentieth century — of a scientific and analytical approach to materials. Réaumur, in the early eighteenth century, was the first modern materials scientist, but his ability to relate traditional practice to advanced science was not followed, for the love of mathematical physics applied to highly simplified models was at that time irresistible and the obscurantism of the alchemists had repelled the disciples of Newton.

Kranzberg and Smith's survey is followed by a section with its eyes wholly on modern times: Richard Claassen and Alan Chynoweth, two industrial scientists, write on "MSE as a Multidiscipline". Specifically, they trace the routes by which physicists, chemists and engineers of various persuasions combined to create the profession of MSE, and they do so primarily through a collection of case histories of technical developments (heatshield design for space vehicles; the developrnent of the transistor; plastic coatings on razor blades; artificial fibres; optical lightguides, and others). These stories, combined with an analysis of the implications for the organisation of interdisciplinary research projects, provide the flavour of this section.

Kranzberg and Smith, the historians, quote the Greek historian Xenophon ("What are called the mechanical arts carry a social stigma and are rightly dishonoured in our cities") and Plutarch ("for it does not of necessity follow that, if the work delight you with its grace, the one who wrought it is worthy of esteem"), to indicate what the ancients thought of craft and skills applied to materials. Of all the ancient exponents of the 'mechanical arts', the slave miners of antiquity were perhaps the most abjectly oppressed, miserable and shortlived.

Today, once again, 'mechanical artists', and miners in particular, are apt to be regarded by wielders of temporal power as the carriers of social stigma. This surprising conclusion emerges with great clarity from another recent publication, the proceedings of the 1978 American Mining Symposium, the subtitle of which was "Politics, Society and the Mineral Industry". The proceedings are published in the journal *Materials and Society* [2]. Several of the lectures are germane to the matter in hand: these are written by a metallurgist, a congressman, a copper mining expert, a coal mining expert and J. Allen Overton, the president of the American Mining Congress, who finds himself under the necessity of acting as a fulltime lobbyist. Their complaint — the word is too weak, it is a succession of cries of despair — is the unchecked growth of regulation and of the body of state employees who enforce regulations and steadily add to their number. The chairman remarked that the "political outcry", as he termed it, at the symposium was unpremeditated and uncolluded and the more credible for it. The tone is exemplified by two extracts from Allen Overton's lecture: "Time and again, we see no attempts to consider trade-offs or weigh the cost of regulations against their supposed benefits. This is the work of zealots, and the only guideline they know is their own self-proclaimed virtue. The repeated failure — I should call it the obstinate refusal — to take acount of the economic consequences mean that some regulations can have ruinous impact". And again: "Finally, there is bewilderment and foreboding about the sheer mass of regulations, their frightening complexity, the paralysing uncertainty about what the bureaucrats will propose tomorrow, and how the rules will be changed in the middle of the game". The changing of rules in the middle of the game is exemplified by William Simon, a former US Secretary of the Treasury, in another recent publication[3]: "Manufacturers of children's sleepwear were forced by the government to process children's sleepwear with a flame-retardant chemical. When the companies shifted over to the costly new process, the FDA banned the flame-retardant chemical as a suspected carcinogen, and the manufacturers were ordered to recall their merchandise and to compensate buyers".

In such an atmosphere, it becomes understandable why American oil policy is in such disarray: those who fully understand the situation in prospect plead for decontrol and a partial release from the burden of regulation (McKetta's lecture on American energy policy at the symposium is a case in point), but their word is widely distrusted by politicians precisely because they are experts. Xenophon stalks among us again. The proponents of regulation, and those who support and encourage them, are all tooofen convinced that experts are venal and devoid of public spirit (just as the free citizens of ancient Greece regarded their bonded craftsmen). When the mining specialists plead public interest, the fact that the plea comes from them at once invalidates it. But the inevitable consequence is that the same judgment is being applied to the regulators themselves: they have by now as great an interest vested in the increase of regulations as the protesters have in their abatement. Not only that: those miners, MSEs, and may others to whom by implication Xenophon's cold dictum is attached, conclude that their honesty is disbelieved by some of the most powerful in the land, and when a professional's integrity is systematically impugned, he is driven at last towards political extremes; at the least he will come to distrust the regulators' own honesty of purpose. The prospect is not enticing, and the American crisis of regulations should give us in Britain pause.

1. *Materials Science: Its Evolution, Practice and Prospects* (Elsevier-Sequoia, Lausanne, March 1979).

2. Proceedings of the 1978 American Mining Symposium, *Materials and Society* **3**, 1 (1979).

3. William Simon, *A Time for Truth* (McGraw-Hill, New York, 1978).

80

Differential Attack

Nature, **282**, 555 (1979)

The essentially empirical science of chemical etching, used as a routine tool by the material scientist to reveal the microstructure of mineral samples, has developed in some unexpectedly sophisticated ways. One is the recent development of a thin-film ultramicrosieve[1], the latest in a line of filters whose unlikely origins lie in an observation made years ago that chemical etching could reveal the damage tracks left by the passage of nuclear fission fragments through thin mica sheets.

This story begins in 1961, when R.M. Walker and P.B. Price at the Central General Electric Research Laboratory in Schenectady discovered that these damage tracks were revealed by chemical etching which pierced minute, uniformly-sized holes through the mica. Other minerals were soon found to behave similarly, and other forms of radiation found sometimes to yield etchable tracks. Walker and Price, aided by R.L. Fleischer, over the next few years applied this differential etching method to rnany problems in geochronology, planetary physics (the study of meteorites in particular), palaeontology and nuclear physics. The field has become large enough to achieve that ultimate accolade, a journal of its own.

The early mica work came to the attention of S.H. Seal, a clinical cancer researcher at the Sloan-Kettering Institute who needed an ultrafilter to hold back cancer cells from suspension in blood; nothing he had tried would do the trick. Price's irradiated and etched mica foils were just right (the holes were all 3-4 μm across) but the foils were too fragile. At this point, Fleischer found that a polycarbonate polymer could be substituted for mica and had better mechanical properties. Thus began the manufacture of General Electric's *Nucleopore* biological filters, which among other uses have a role in cancer diagnosis. The filters were described by Fleischer, Price and Symes in *Science* [2] and the steps in this weird progression from fission damage research to clinical instrumentation are traced, step by step, in an excellent survey, *Applied*

Science and Technological Progress .

This case-history has evidently influenced subsequent research at General Electric. For instance, W. Desorbo and H.E. Cline, who were investigating directionally solidified eutectics, consisting of parallel metallic rods in a matrix, with a view to their use in jet engine turbine blades, found that differential etching left the rods protruding clear of the matrix. Porous metallic replicas made from such etched samples performed neatly as microsieves[3].

Filters with holes a few microns across are considered coarse nowadays (too coarse, for example, for the isolation of airborne smokes), and a new approach has now shown how to make filters with sub-micron apertures. M. Kitada[1] co-evaporated gold and germanium in vacuum onto sodiurn chloride crystals. Films 50–500 nm thick were evaporated, at or near the eutectic composition of 12 at % Ge. Germanium was deposited as round grains 20-250 nm across, depending on alloy composition and substrate temperature. The films were floated off in water and attacked by an iodated mixed-acid etchant, which selectively removed the more reactive germanium. Since the germanium grains grow in columnar fashion through the film thickness, removal of germanium leaves minute pores right through the residual gold films and the result is a thin-filrn ultramicrosieve.

Another form of selective attack which is not only technologically valuable but scientifically intriguing, is the vertical etching of silicon. This technique, which for several years has been quietly developed within the semiconductor industry, is now beginning to find applications in other scientific fields. An excellent account of the technique, its scientific basis and recent very various applications has just been published by D. L. Kendall[4] who regards it as a "whole new way of structuring and using a solid material". The essential fact is that extremely narrow, deep grooves with depth-to-width ratios exceeding 100:1 can be etched in silicon. It is only necessary to start with a monocrystal slice accurately parallel to (110), and to prepare an oxide surface mask with narrow gaps precisely parallel to the (110)/(111) intersection. Etching is very much faster in the [110] than in the [111] direction, because a newly formed (111) plane is quickly covered with an oxide film which severely hinders access of etchant (usually an aqueous solution of potassium hydroxide used at 80°C), whereas other planes are not thus affected. Any divergence of the surface slit from correct alignment enhances the rate of etching of the sides of the groove.

Whereas it is easy to etch deep ultra-narrow grooves, it is not possible to etch round holes. A regular array of exact 5 μm- square holes has been fabricated by etching two populations of grooves, mutually inclined at 70°, from opposite sides of a (110) slice until they just meet[5].

Other uses include deeply etched solar energy absorbers (effectively thin-layer black bodies), an elaborate high-area Si/SiO_2 capacitor only about 1 mm square, diffraction gratings efficient for high-order optical spectra and an infrared reflection polariser. The most subtle application is a device for ejecting ink drops between closely spaced pairs of electrified p/n junctions to charge the drops, which can subsequently be electrically steered to form a printed image. This was cleared the way to computer-controlled printing without moving parts.

A variant on the sieve theme is a selective-ion device for separating ions such as Cu^+, which diffuse very fast in silicon. A thin-walled meander is formed in a silicon slice by etching offset parallel grooves from the two sides; copper will pass selectively through such a silicon membrane, much as hydrogen passes selectively through palladium. If the membrane is oxidised by anodising in hydrofluoric acid, the oxide has pores which can be as small as 1 nm in diameter. These can be used to separate molecules of different sizes: larger pores, for example, can serve as virus filters[6,7].

1. Kitada, *J. Mater. Sci.* **14**, 2765 (1979).
2. Fleischer, Price and Symes, *Science* **143**, 249 (1964).
3. DeSorbo and Cline, *J. Appl. Phys.* **41**, 2099 (1970).
4. Kendall, *Annual Reviews of Materials Science* **9**, 373 (1979).
5. Bean, *IEEE Trans.* **ED-25**, 1185 (1978).
6. Watanabe *et al.*, *J. Electrochem. Soc.* **122**, 1351 (1975).
7. Gregor and Gregor, *Sci. Amer.* **239**, 88 (1978).

81

A Soft, Superionically Conducting Lower Mantle?

Nature, **308**, 493 (1984)

The meagre contribution so far of materials science to the general understanding of the Earth's mantle is about to be enriched, in part by the study of ionic crystals whose properties may be a guide to those of the mantle rocks, with their capacity to undergo plastic deformation and with their large electrical conductivity.

Two lines of geophysical arguments have helped to define what needs to be explained. First, studies of *post-glacial rebound*, the gradual rising of the Earth's surface after the removal of overlying ice sheets, is a pointer to the viscosity of the convective mantle. Some, however, propose a uniform viscosity for the whole mantle, others, layered viscosity variations. Weertman[1] takes it as probable that the lower mantle is orders of magnitude more viscous than the mean mantle viscosity.

Second, the conductivity of the mantle, especially at its lower levels, is linked with the delays between changes in the magnetic field of the core and its manifestation at the Earth's surface — what magneticians call magnetic viscosity. Thus Anderson[2] has recently pointed out that the delay associated with a magnetic 'jerk' such as that of 1969 must be intimately linked with mantle conductivity, but has also drawn attention to the difficulties in putting a value on the mean conductivity from such observations. Nevertheless, Stacey[3] has estimated, from such data together with an analysis of the secular variation of the length of the day, an electrical conductivity at the bottom of the mantle (itself orders of magnitude higher than in the upper mantle) of 100–500 $\Omega^{-1}m^{-1}$.

It is widely held that the major constituent of the lower mantle is a pressure-induced polymorph of pyroxene, $(Mg,Fe)SiO_3$, with perovskite structure. The implication is that the mantle is chemically but not structurally homogeneous (though Whaler[4] has recently advanced seismic arguments for presuming chemical heterogeneity).

In assessing the compatibility of the conventional model of the lower mantle with the values of viscosity and conductivity needed to fit data on glacial rebound and magnetic delay, the difficulty arises that (Mg,Fe)SiO_3 with perovskite structure can be made in the laboratory only at such high pressures, and therefore minute volumes, that direct measurements of its properties are not feasible.

In any case, in spite of the large numbers of perovskite-type compounds known[5], there has been no investigation whatever of the plastic deformation of any of them. Indeed, Poirier *et al.* [6] point out that there is no direct knowledge of the phase (*pace* Whaler) presumed to occupy 40 per cent of the Earth's volume, and they have accordingly undertaken a study of $KZnF_3$, a salt with perovskite structure which is stable at ambient pressure, up to its melting point of 870°C. In this they followed the lead of O'Keeffe and Bovin[7], who examined $NaMgF_3$, another perovskitic salt, and found that it becomes a solid electrolyte (alias a *superionic conductor*) at high temperatures.

The special importance of the work of Poirier and his colleagues is that they examined the mechanical as well as electrical behaviour of $KZnF_3$, showing both that it becomes a superionic conductor and that it acquires a low viscosity above the same critical temperature range. This remarkable finding is of importance both for geophysicists and for materials scientists.

Specifically, Poirier *et al.* have measured the creep curves of oriented crystal slices of $KZnF_3$ between 650° and 844°C and electrical conductivity between 650° and 800°C. Both primary and steady-state creep was observed. The latter has approximately newtonian characteristics, with a rate proportional to stress at a fixed temperature. At a fixed stress, the creep rate increases with temperature up to 750°C, decreases slightly between there and 800°C and then increases steeply at higher temperatures.

The conductivity rises slowly with temperature up to a value of $0.8 \pm 0.2 \times 10^{-2}\ \Omega^{-1}m^{-1}$ at 700°C and then rapidly to $1.9 \pm 1.8\ \Omega^{-1}m^{-1}$ at 800°C, with most of the increase between 700° and 750°C. The values agree well with those reported for $NaMgF_3$ by O'Keeffe and Bovin — but the high degree of experimental scatter near 800°C should be noted.

Weertman[1] had already concluded that newtonian creep is necessary to account for observations of post-glacial rebound. Newtonian creep usually involves diffusional (Nabarro) creep in fine-grained materials, not in single crystals as in Poirier's work. Poirier *et al.* conclude that they have observed an example of Harper-Dorn creep, a mechanism reported many years ago but badly understood. They argue that the sharp increase of diffusivity of one ionic species, here F^-, associated with the advent of superionic conduction (see for example Betancourt *et al.* [8]) should hinder creep by promoting the climb and immobilization of some dislocation segments in crystals of $KZnF_3$. Consequently, dislocations should not multiply in the familiar way during plastic deformation, but pre-existing dislocations should drift under stress according to the Orowan law, which specifies a drift rate proportional to stress.

Poirier *et al.* [1] also discuss in some detail possible mechanisms for the hardening of the crystals between 700° and 750°C. They further point out that, once the crystals become superionically conducting, the F^- sublattice effectively 'melts'[9] and this entails a change in the core structure of dislocations to a B2,β-brass type, with resultant easy slip on (100) as observed.

From a geophysical standpoint, if perovskitic $(Mg,Fe)SiO_3$ behaves similarly to $KZnF_3$, and if it is is indeed the major constituent of the lower mantle, then the viscosity of the mantle would not increase with depth, as has hitherto been assumed, but might actually decrease below ≈ 700 km. The maximum conductivity of $KZnF_3$, about $2\ \Omega^{-1}m^{-1}$, is still about two orders of magnitude less than the value deduced as consistent with superionic conductivity in perovskitic $(Mg,Fe)SiO_3$. But as the maximum conductivities of known ionic conductors range up to $500\ \Omega^{-1}m^{-1}$ this disparity poses no problem.

To materials scientists, the interesting feature of these measurements is the occurrence, in a narrow temperature range, of steep changes in both viscosity and conductivity. A check through two major conference proceedings concerned with superionic conductors[8,10] reveals no investigations whatsoever of the mechanical characteristics of any member of this family of materials. There is, however, another family of crystals which shares with superionic conductors the characteristic that one sublattice only undergoes progressive disordering as the temperature is increased. These are the *plastic crystals,* organic compounds in which some of the molecules acquire both positional and orientational disorder and which are so called because, in the partially disordered state, they become extremely soft so that some will flow even under gravity. When that happens, self-diffusivity also becomes extremely high. and it is widely assumed that the two changes are linked, but the nature of the relationship is not clear. The general characteristics of positional and orientational disorder in many types of crystal, including ionic superconductors, and of the flow and self-diffusion of plastic crystals are all very clearly set out in an excellent monograph by Parsonage and Staveley[11]. While it is not likely that plastic crystals (molecular) and superionic conductors (ionic) have similar mechanical characteristics, the work of Poirier and his colleagues has nevertheless opened an entirely new field of research into the relation between the disordering or 'melting' of one sublattice in a crystal and the consequential changes in diffusivity of either ions or molecules, and the further connection between these two features of the crystal and its mechanical softening.

1. J. Weertman, *Phil. Trans. R. Soc. (Lond.) A* **288,** 9 (1978).
2. D.L. Anderson, *Nature* **307,** 114 (1984).
3. F.D. Stacey, *Physics of the Earth* (Wiley, New ¥ork, 1969).
4. K.A. Whaler, *Nature* **306,** 117 (1983).
5. F.S. Galasso, *Structure, Properties and Preparation of Perovskite-type Compounds* (Pergamon Press, Oxford, 1969).
6. J.P. Poirier, J. Peyronneau, J.Y. Gesland and G. Brébec, *Phys. Earth Planet. Interiors* **32,** 273 (1983).
7. M. O'Keeffe and J.O. Bovin, *Science* **206,** 599 (1979).
8. M. Betancourt *et al.,* in *Fast Ion Transport in Solids — Solid State Batteries and Devices* (ed. by W. van Gool) 233 (North-Holland, Amsterdam, 1973).
9. H.A. Schulz, *Rev. Mater. Sci.* **12,** 351 (1982).
10. P. Vashishta, J.N. Murphy and G.K. Sheney, (editors) *Fast Ion Transport in Solids — Electrodes and Electrolytes* (North-Holland, Amsterdam, 1973).
11. N.G. Parsonage and L.A.K. Staveley, *Disorder in Crystals* , 240, 512, 696 (Clarendon Press, Oxford, 1978). — See also article 87, below.

82

Limits to Coulomb's Law

Nature, **319**, 177 (1986)

Many of the scientific laws that the readers of this journal absorbed in their youth are more than a century old and a surprising number go back more than two centuries — but which of them can still claim exact congruence with experimental facts? Some continue in daily use as sound approximations, for example, Boyle's, Hooke's and Ohm's laws, but lose their applicability in specific contexts, such as the behaviour of supercritical steam, 'potty putty' or semiconductor resistivity. Others, such as Newton's laws of motion, remain very close approximations to the behaviour of the real world until extreme conditions (for example, velocities) are attained. In this issue of *Nature* [1], K. Kendall challenges another venerable law — Coulomb's law, formulated in 1773, which describes the frictional behaviour of a bed of powder.

According to Coulomb's law, the frictional force between a particle and a smooth plane pressed together by a normal force w is expressed as $f = k + \mu w$, where μ is the friction coefficient and k is a constant representing the intrinsic interaction — the van der Waals attraction — between the mating surfaces. Coulomb's law has been used for a long time in engineering calculations, particularly in soil mechanics. Kendall now reports on experiments with wet sand confined between two silica plates and finds that the friction coefficient varies substantially, both with normal load and with particle size.

The new findings imply that the van der Waals interaction between, for example, two equal spheres depends on their diameters and on the force pressing themtogether; this was first established by Johnson, Kendall and Roberts[2]. If other factors remain constant, the friction in a powder bed will rise rapidly as finer particles are used. In view of the widespread use of ultrafine powders in advanced ceramics, this finding has considerable practical importance. As with other named laws, Coulomb's law will presumably remain an adequate approximation under the normal circumstances of coarse powders and soil mechanics, but not under extreme conditions.

Coulomb's law is just one small feature of the varied, untidy but widely useful field of powder mechanics, a term not commonly used, but one that collectively covers dynamics, statics and thermodynamics. Consider some of the ways a powder can behave. When dry and in a still atmos phere, a powder has a 'flowability' that depends on particle mean size. size distribution, mean shape and shape distribution and on intrinsic frictional properties. An assembly of needles may stick obstinately, whereas dry, uniform-sized sand flows easily. The capillary effect of moisture iscrucial: common salt needs a hygroscopic additive to keep it free-running; dry powder snow responds to pressure quite differently from wet snow, as any skier knows. If the proportion of water becomes too high, the result may be an avalanche or a quicksand.

A dry powder can be made fluid by blowing air or some other gas through it. Such a 'fluidized bed' has much lower friction than a normal powder bed and resembles a highly mobile liquid. The theory of fluidized bed mechanics and heat transfer is important in chemical engineering, because such beds are commonly used for reacting solids with hot gases. When the gas pressure becomes too high, the agitation of the powder leads to erosion of the container. Such erosion has implications ranging from particle impact on the windscreens of supersonic aircraft to the sand-blastling of castings or spark plugs. The abrasive wear of surfaces by powders ha salso become a large field of study.

The optimum feasible irregular packing of powders leads to complex statistical issues, which have been treated by Bernal and Finney[3]. Their concept of dense random packing of equal-sized spheres, originally conceived as a model for monatomic liquids, has been extremely influential in guiding the modelling of metallic glasses. Surprisingly, a randomly packed assembly of monosized spheres can be within about five percent of the density of a regular close-packed array. Dense random-packed arrays can be further densified by sintering, a technique that goes back to prehistoric practice in the baking of pottery. The stereology of solid-state sintering — the developing relation between packing fraction, porosity, permeability and specific surface area — has become a major field of scientific enquiry with relevance to metallurgy, ceramics and catalysis, and several novel experimental and conceptual tools have evolved with it. The mechanism of sintering — notably the crucial role of the curvatures of internal free surfaces and of intercrystalline boundaries — is now well understood. Through doping, subtle control of the motion of such boundaries has been achieved, making it possible to sinter to full density; the absence of scattering from pores has enabled the creation of translucent or transparent polycrystalline ceramics, now used in efficient high-temperature lamps.

The ubiquity of powders in applied science indicates that Kendall's discovery of a major departure from a long-accepted powder law is of wider import than might at first appear. We still have to see how representative the circumstances of Kendall's experiments are. He used a wet charge and the moisture fraction is not stated; presumably dry sand would behave quite differently. The plastic domain, as in hot isostatic pressing of metallic powder compacts, must entail frictional regimes quite different from those in the elastic regime treated by Kendall. Even so, his work has reopened a field that has been hemmed in by orthodoxy and neglected because ultrafine powders have only recently become commercially important.

1. K. Kendall, *Nature* **319**, 203 (1986)
2. K.L. Johnson, K. Kendall and A.D. Roberts, *Proc. R. Soc. (Lond.O A* **324**, 301 (1971).
3. J.D. Bernal and J.L. Finney, *Nature* **213**, 1079 (1967); *Proc. R. Soc. (Lond.) A* **319**, 479 (1970).

83

Political Science

MRS Bulletin, **16/1**, 72 (January 1991)

It is not often that a scientist or engineer attains high public office. Herbert Hoover, a mining engineer; John Sununu, a nuclear engineer; Margaret Thatcher, a chemist turned lawyer... all were highly unconventional in their careers. More often, we find scientists and near-scientists engaging in political activity: the priorities of such as Noam Chomsky are common enough not to occasion surprise today. Margaret Thatcher does not often need to consult chemical principle to guide her political decisions (environmental matters apart); Chomsky, when fulminating against Vietnam policy could hardly have been moved by philological passion.

A real danger appears, however, when an active professional scientist gains political power as well, and uses that to proselytize, or even impose a particular scientific theory. One such person was Marcelin Berthelot (1827-1907), an eminent French chemist who did not believe in atoms. Berthelot remained active in research almost to the end of his days — he was a man of demonic energy — but he also became, for a while, Inspector of Higher Education and even Foreign Minister in the French government. He was regularly elected president of the French Chemical Society as well.

His scepticism concerning atoms arose, to simplify the matter, from a conviction that the 'atomic hypothesis' required all gaseous molecules to be diatomic, and since this led to contradictions, the entire hypothesis must be erroneous. Because of the highly centralized nature of French education (university professors still, nominally, require ministerial approval for a change in course structure and content), and because of the excessive power of a *grand patron* in those days, Berthelot apparently succeeded in indoctrinating generations of French chemistry students with his obsession.

If there were no atoms, then stereochemistry had to be a fantasy (in spite of Pasteur's towering achievements!) and without that intellectual tool, organic chemistry could not develop properly. This recognition led the French chemist Haller in 1900, the year when Berthelot was elected to the French Academy, to declaim furiously that Berthelot's piece of "pure doctrine" had disastrously held back the production of organic chemicals in France, while it roared ahead in Germany. Organic chemistry, Haller declared, is "directly inspired by theory" and a false theory meant mis-inspiration. One recent commentator has gone so far as to claim that young French chemists continued to be taught atomic scepticism until the eve of the atomic bomb!

The only safe vaccine against episodes of this sort is a multiplicity of centers of initiative... free and entirely independent universities, multiple independent professional bodies (preferably in competition), and multiple journals in the same field. Today, science is international to a degree that protects against national extravagance... unless, of course, frontiers are closed; then the Lysenkos of this world seize their chance. The most important protection, perhaps, is a multiplicity of journals: in a country where there is only one official journal in each field and the international mails are restricted, both standards and liberty are apt to slide and error can take long to resolve (as happened in the polywater episode). MRS has done well to create *Journal of Materials Research* , and should not be fazed by possible criticisms that it has diluted the literature!

If a scientist can misuse political influence, political dogma can, on occasion, mislead a scientist (as it did Lysenko). The most amusing example of this (of which I have personal experience) refers to Le Chatelier's principle. This was enunciated by another very eminent French scholar, a metallurgical chemist this time: he invented the Pt/Pt-Rh thermocouple which is still in use a century later. We all learn his principle in school: a skater glides over an ice sheet at 0°C, the pressure he exerts lowers the freezing point and the skate sinks a little deeper into the locally melting ice. The system has "accommodated" the applied constraint... that's how Le Chatelier worded his thermodynamically based Principle.

During the Second World War, the supply of natural quartz crystals from Brazil almost dried up. They were needed for oscillators, and so large crystals were artificially grown by the hydrothermal route. The crystal thus produced contained Dauphiné twins, regions in which the lattice is rotated by 180° about the trigonal symmetry axis. Such crystals were useless unless the twinned regions could somehow be rotated back into the surrounding orientation, restoring a true monocrystal.

An eminent Cambridge crystal physicist, W.A. Wooster, found out how to do this: quartz is elastically highly anisotropic and so, if a stress is applied to a twinned quartz crystal, the differently oriented inclusion will store a greater or lesser amount of elastic energy (per unit volume) than the main orientation. The trick, as Wooster recognized, is so to arrange the stress system that the desired orientation (the main crystal) is elastically 'softer' in response to that constraint, and at the same time to raise the temperature so that atoms can easily shift through short distances. That orientation which gives way more compliantly — which "accommodates the constraint" — will prosper; the other orientation will shrink and disappear. By applying carefully calculated torques and bending moments to crystal plates that were to be used in oscillators, Wooster and his colleague L.A. Thomas were able to remove the trouble- some imperfections. They published their findings later, in 1951.

At about the same time, a Soviet crystallographer, A. V. Stepanov, treated the same problem in a series of theoretical papers. Unfortunately, he stood Le Chatelier's Principle on its head. He asserted that the crystal must modify its local orientation in such a way that the system maximizes its *resistance* to the constraint — clearly, a distinctly communist world-view! The final irony in this little history is that Wooster himself was a dedicated communist, quite open about his predilections and always ready to defend the Soviet position. But for this excellent crystal physicist, scientific rigour quite unconsciously outweighed the commu nist way of looking at the world.

84

Metallic Solid Silicon

Nature, **357**, 645 (1992)

"Silicon is the most intensely studied material in the scientific literature". While this recent assertion[1] is undoubtedly true, remarkable new aspects of silicon are nevertheless still being discovered. One such is a transition from the normal semiconducting (non-metallic) state to a metallic state in the neighbourhood of tiny plastic indentations on pure silicon crystals, created by a diamond pyramid under load[2-6]. This discovery, the final outcome of a long series of theoretical insights in electron theory and experiments initially in high-pressure laboratories, has about it the unmistakable scent of a major advance: it is likely to lead to a host of consequential researches.

The theoretical insights referred to go back to Mott's treatment[7] of the metal–nonmetal (MNM) transition in group IV semiconductors. This transition, from a semiconductor to a metallic state, was known to depend upon doping the semiconductor (Si or Ge) with enough pentavalent solute so that the localised atomic wave functions of the solute atoms begin to overlap, screening the extra electrons from the extra nuclear charge of the solute atoms. Mott's celebrated quantitative criterion for the MNM transition of a doped semiconductor (or insulator), as well as several later variants of the thoery, are particularly clearly explained by Cottrell in a recent book[8], and the remarkable consistency of the theory with a very wide range of experimental findings has been recorded by Edwards and Sienko[9].

The original theory referred only to MNM transitions generated by doping, but a later version, the 'polarization catastrophe' picture based on pre-quantum ideas due to Goldhammer and Herzfeld and developed by Edwards and Sienko[9,10], shows that an MNM transition can equally be forced by compressing a semiconductor, increasing the atomic density until the dielectric constant diverges (a 'cooperative polarization') at a critical density. Cottrell presents this clearly also.

It is well established that ordinary "diamond-cubic" silicon transforms to the much denser β-tin structure under hydrostatic pressure at ambient temperature; the critical pressure (A) is in the range 11.3–12.5 GPa[11]. This is close to the indentation hardness (B) of diamond-cubic silicon, which is in the range 11-12 GPa. (Later, it emerged that in fact B slightly exceeds A). Several physicists were struck by this similarity and suggested (e.g., Gerk and Tabor[12]) that silicon might transform to the β-tin structure under a diamond indenter, where the stress state is a combination of hydrostatic and deviatoric components. This suggestion was consistent with an earlier experiment, in 1972, by Gridneva *et al.* [13], who found a large, reversible drop in resistivity in a thin layer surrounding an indentation in silicon, as would be expected in view of the metallic character of the β-tin structure. The recognition of the close similarity of A and B and the Russian experiment[13] underlay some of the recent rash of new experiments... but serendipity played its customary role as well.

The first study, mostly carried out at IBM's Yorktown Heights laboratory, was by Clarke *et al.* [2]. They observed the formation of an *amorphous* phase under a diamond pyramid pressed into a silicon crystal. This research was part of a programme to study crack formation in brittle solids. Once the unexpected phase transformation had been observed, the experimenters recalled Gerk and Tabor's suggestion and measured the electrical resistivity: they found a reversible reduction near the indentation.

Clarke next joined forces with Pharr, in Texas, and Oliver, in Tennessee[3]. Oliver had invented an instrument, the *Nanoindenter*,which allowed accurate hardness measurement to be made under exceedingly small loads[14]. As part of a broad study of nanoindentation hardness of many materials, silicon crystals of various orientations were studied. Two features unique to silicon were found: with (relatively) large peak loads, a sharp discontinuity in displacement was found during the *un* loading cycle; with smaller peak loads, a reversible hysteresis cycle was observed. The discontinuity was attributed to cracking (confirmed by direct experiments two years later); the hysteresis (which was accompanied by a large resistivity change) was associated with a pressure-induced phase transition.

The next step was the observation[4] of unmistakable plastic extrusions around nanoindentations in silicon (Fig. 1). This, again, was unique to silicon among the many brittle materials examined. The implication is that silicon transforms to a metastable form under the diamond and that the plastic behaviour, as well as the electrical properties, become those characteristic of a metal.

This preliminary study was followed by a more detailed one, carried out cooperatively by a Texas/Tennessee/New York/California consortium (Clarke had in the meantime moved to the University of

California)[5]. The mechanical and electrical hysteresis were thoroughly analysed, by depositing an array of strip electrodes on the surface and indenting on or between the strips. The effects due to the MNM transition

Fig. 1. Scanning electron micrograph of a 40-millinewton indentation in silicon. (Courtesy of G.M. Pharr).

were only seen if an indentation pierced or touched an electrode (Fig. 2, next page). The authors emphasize that the resistivity change is due partly to the phase transition and partly to changes in the characteristics of the electrode/silicon interface as the transition reaches each electrode: the Schottky barrier is replaced by an ohmic, low-resistive metal-to-metal interface.

Independently and very recently, Gilman (who has a long record of research on 'electromechanical' properties) reviewed[15] a large number of MNM transitions in a range of semiconductors and insulators and

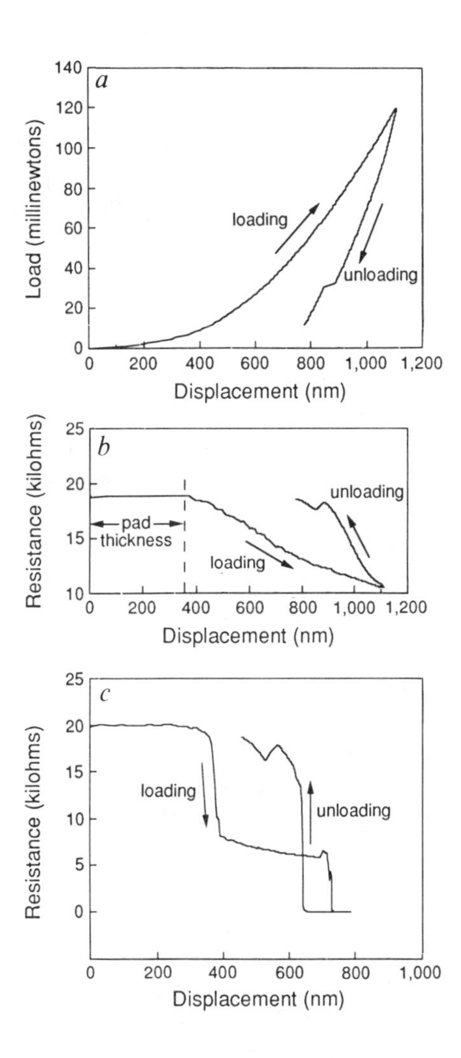

Fig. 2. Indentation of silicon. (a) Displacement hysteresis during a loading-unloading cycle and the strain discontinuity during unloading. (b) Resistivity change for an indentation *through* a gold electrode; the change is small because much of the current path is through unmodified silicon between adjacent electrodes. (c) Resistivity changes when the indenter bridges a gap between electrodes, so that the amorphized silicon touches both electrodes.

showed some striking correlations between the experimental 'metallization pressures' and values calculated from the polarization catastrophe model. He went on to predict that many such transitions should be observable under indenters, and speculated whether plastic shear may play a role in the MNM transition. He further suggested that, alternatively, 'metallization' and dislocation motion may both be correlated with the excitation of bonding electrons into antibonding states, and offers some evidence from the literature in support of this idea.

In the meantime, Pirouz *et al.* [16] have been conducting a separate series of experiments to study the partial transformation of diamond-cubic silicon to a "diamond-hexagonal" form by indentation at 400-650°C. (The experiments by Clarke, Pharr, Oliver and colleague were at room temperature). Pirouz *et al.* showed that this process is martensitic (shear-induced); here the deviatoric stress component under an indenter is crucial. But the authors have apparently not looked for an MNM transition in the hexagonal silicon.

There remains the question of how the amorphous, conducting phase[2] is formed. Conventional pressurization experiments have hitherto shown no sign of it. This is addressed in Pharr's latest contribution[6]. He marshals the extensive published evidence to show that silica, ice and several semiconductors other than silicon become amorphous under pressure[17]. Indeed, non-hydrostatic stresses, at the edge of an indentation, help give silica two distinct pressure-induced amorphous phases. as the editor of *Nature* himself emphasized some years ago[18]. It is also clear from earlier experiments that thin amorphous silicon films can undergo an abrupt MNM transition under hydrostatic pressure[19] (just as unpressurised amorphous silicon 'melts' to produce a denser, metallic melt — see a summary by Cahn[20]). Pharr finally concludes that "at some point in the transformation, an amorphous phase is formed". As people are wont to say in grant proposals, more research is clearly needed!

1. H.R. Huff, in *Concise Encyclopedia of Semiconducting Materials and Related technologies* (edited by S. Mahajan and L.C. Kimerling) 478 (Pergamon Press, Oxford, 1992).
2. D.R. Clarke, M.C. Kroll, P.D. Kirchner, R.F. Cook and B.J. Hockey, *Phys. Rev. Lett.* **60**, 2156 (1988).
3. G.M. Pharr, W.C. Oliver and D.R. Clarke, *J. Electronic Mater.* **19**, 881 (1990).
4. G.M. Pharr, W.C. Oliver and D.S.Harding, *J. Mater. Res.* **6**, 1129 (1991).

5. G.M.Pharr, W.C. Oliver, R.F. Cook, P.D. Kirchner, M.C. Kroll, T.R. Dinger and D.R. Clarke, *J. Mater. Res.* **7**, 961 (1992).

6. G.M. Pharr, in *Thin Films: Stresses and Mechanical Properties III* (eds. W.D. Nix*et al.*), Mater. Res. Soc. Symp. Proc. **239** (Materials Research Society, Pittburgh, 1992), in press.

7. N.F. Mott, *Proc. Phys. Soc. A* **62**, 416 (1949).

8. Alan Cottrell, *An Introduction to the Modern Theory of Metals* 16 (The Institute of Metals, London, 1988).

9. P.P. Edwardsand M.J. Sienko, *Phys. Rev. B* **17**, 2575 (1978).

10. P.P. Edwards, and M.J. Sienko, *Int. Rev. Phys. Chem* **3**, 83 (1983).

11. J.Z. Hu, L.D. Merkle, C.S. Menoni and I.L. Spain, *Phys. Rev. B* **34**, 4679 (1986).

12. A.P. Gerk and D. Tabor, *Nature* **271**, 732 (1978).

13. I.V. Gridneva, Y.V. Milman and V.I. Triflov, *phys. stat. sol* **14**, 177 (1972).

14. W.C. Oliver and C.J. McHargue,*Thin Solid Films* **161**, 117 (1988).

15. J.J. Gilman, *J. Mater. Res.* **7**, 535-538 (1992).

16. P. Pirouz, R. Chaim, U. Dahmen and K.H. Westmacott, *Acta metall. et mater.* **38**, 313, 323, 329 (1990).

17. Y.K. Vohra, H. Xia and A.L. Ruoff, *Appl. Phys. Lett.* **57**, 2666 (1990).

18. J. Maddox, *Nature* **326**, 823 (1987).

19. O. Shimomura, S. Minomura, N. Sakai, K. Asaumi, K. Tamura, J. Fukushima and H. Endo, *Phils. Mag.* **29**, 547 (1974).

20. R.W. Cahn, in *Glasses and Amorphous Materials* (edited by J. Zarzycki) 517 (VCH, Weinheim, 1991).

Book Reviews

85

Pirotechnia

Technology and Culture , **8**, 88 (1967)

Pirotechnia.. By Vannoccio Biringuccio. Translated by Cyril Stanley Smith and Maria Teach Gnudi; biographical and bibliographical introductions by C. S. Smith; translator's notes by M. T. Gnudi. Cambridge, Mass., M.I.T. Press, 1966. Pp. xxvii + 477. $3.45 (paperbound).

This translation of the Western world's first metallurgical classic has been photographically copied from the luxury edition first published in 1942 by the American Institute of Mining and Metallurgical Engineers. It is a monument to the percipience and pertinacity of Dr. Smith, a metallurgist by profession (now Institute Professor at M.I.T. and for many years past a rare blend of technologist, scientist, and historian). With Dr. Gnudi's expert linguistic assistance, Smith presented to his professional colleagues a book that not only offers a *first-hand* description in extenso of all the principal techniques involved in the smelting, melting, distilling, casting, and alloying of metals but also a range of shrewd comments by Biringuccio on the technological philosophy of his day, illuminated by a number of more general reflections on life.

According to Smith's biographical essay, Biringuccio was born in Siena in 1480. The stormy politics of that republic obliged him to spend prolonged periods in other parts of Italy and abroad, while the periods of power enjoyed by his princely protector enabled Biringuccio to exercise the metallurgical skills learned during his travels. He distilled a lifetime's knowledge into his book, which was published posthumously in Venice in 1540 and reprinted many times in a number of editions. The book was illustrated by numerous woodcuts, which are reproduced in the translation under review.

The most striking aspect of Biringuccio's book is its determined empiricism. He writes of what he has seen performed and very often of operations that he has carried out or supervised himself. There is little superstition and no mysticism in his text, and in particular he has no use for a priori arguments. He favors a particular way of building a reverberatory furnace or supporting the core of a gun mold because he has compared it with alternative procedures and judges it to be the most efficacious; if Aristotle ever laid down the law on such a topic he offers no hint of it, and he refuses to be intimidated by contrary assertions by other writers if these run contrary to his experience. Occasionally he does retail second-hand information but is at pains to say that he is doing so and adopts a studiously neutral tone on these occasions.

As Smith has pointed out, Biringuccio speculates very little on intrinsic mechanisms. Chemical theory was not at a level that would have allowed him to speculate fruitfully, and in any case he was above all a practical man, concerned to achieve specific material ends. Very occasionally he has recourse to the pseudo–explanations that were the common code of his period: thus he says of a mold that "clay is cold in its own nature, and it not only becomes worse but its coldness is increased by the water that is put in. It is therefore necessary to free it as much as possible from these things if you wish to enjoy the fruits of your labors." However, such passages are notable for their rarity. He has no use for the turgidities of his alchemical colleagues and no sympathy for their aim of converting base metals into gold, though he expresses himself cautiously about the intangible goals of some alchemists. His outlook is perhaps best exemplified by the following quotation: "I know that if perchance some who are passionately devoted to this art should read my work, they would denounce me, accusing me of ignorance and presumption — and if I should hear them, I would patiently agree so as not to quarrel with them. But even if they are more intelligent than I am in this, I do not envy them the blessedness of their knowledge."[1]

While Biringuccio's approach to metallurgy is prescientific then, it is tempered throughout with the scepticism and empiricism that were the precondition of the scientific revolution of the following century. He was plainly convinced of the unity of knowledge and the value of curiosity, and indeed he was not disposed to flaunt his command of technical "secrets" (when very occasionally he summarizes other men's "secrets", the level of his text deteriorates). As he puts it: "I am certain that new information always gives birth in men's minds to new discoveries and so to further information."

Inevitably, the level of his personal expertise varies in different parts of the book. Plainly he was very much *au fait* with furnace construction and with melting and casting, especially of guns. He seems to have been less at home with the intricacies of steel-making (which was most highly advanced in parts of the world that he had no opportunity of visiting). Indeed, writing by hearsay of heathen practice in this connection, he permits himself a touch of superstition when he reports the

Asiatic practice of feeding iron pellets to geese and turning their dung into steel by heating it! As Smith points out, Biringuccio was not clear about the distinction between cast iron and steel. It is, however, worth noting that he is inclined to value the ironsmith above the goldsmith because of the practical value of his work. (It is one of the marks of a society's development beyond the primitive stage that iron loses the taboo that attaches to it in primitive cultures — as attested in the *Golden Bough* — and that its votaries rise in public esteem, even though the religious leaders may lag in this respect.)

In his foreword to this reissue, Smith expresses the hope that this larger edition will reach a wider audience than the 1942/43 edition of 1 500 copies, most of which were snapped up quickly by discerning metallurgists. The volume is not only a major historical source book but entertaining to read for its many philosophical digressions. Indeed, historians of science as well as of technology should find the book stimulating, because in Biringuccio they will recognize one of the immediate precursors of the scientific approach. The book deserves a wide and varied readership.

86

Materials and Man's Needs

Technology and Culture, **17**, 570 (1976)

Materials and Man's Needs: Supplementary Report of the Committee on the Survey of Materials Science and Engineering (COSMAT). Vol. 1: *The History, Scope, and Nature of Materials Science and Engineering.* Vol. 2: *The Needs, Priorities, and Opportunities for Materials Research..* Vol. 3: *The Institutional Framework for Materials Science and Engineering.* Vol. 4: *Aspects of Materials Technology Abroad.* Washington, D.C.: National Academy of Sciences, 1975. Pp. 259; 347; 270; 299. $8.75 (vol. 1); $10.75 (vol. 2); $8.75 (vol. 3); $9.50 (vol. 4). Available from the National Technical Information Service, 5285 Port Royal Road, Springfield, Virginia 22151.

Materials and Man's Needs: Summary Report of COSMAT. Washington, D.C.: National Academy of Sciences, 1974. Pp. 217. $5.75.

Some years ago, the veteran British physicist, E. N. da C. Andrade, concluded a metallurgical lecture with the words: "One blushes, one stammers, one relapses into silence." One might suppose that this merely represented a physicist's natural modesty in the face of metallurgical complexities, but to anyone who knew Andrade, the words lacked a certain measure of verisimilitude; since he was speaking in France, an element of rhetoric was perhaps implicit in his formulation.

In contemplating the remarkable achievement of COSMAT, I am tempted to emulate Andrade's response: comment and criticism stand abashed before such evidence of selfless labour long sustained. But the temptation must be resisted, for the report needs to be drawn to the attention of the wide readership which stands to profit from it. This is not a publication merely for the fraternity of materials specialists; it offers something to historians, politicians, devotees of science policy, and to captains of industry. One can appreciate the point by comparing this report with a book (*Perspectives in Materials Research*), published in 1963 by the U.S. Office of Naval Research, also devoted to an analysis of the then current state and likely trends of research in materials science and engineering. That useful book was purely scientific and technical in conception, whereas the COSMAT report starts with a major incursion into history, and turns its attention again and again to weighty issues of organization, financing and international relations, and to the competing claims of planning and pluralism. In large stretches of the report, materials are left aside: the discussion widens to embrace all of technology. This is entirely as it should be: interdisciplinarity is one of the key concepts in the report, and it is right and proper to draw conclusions about the ostensible theme, materials science and engineering, from more general insights into other fields of science and engineering, and indeed into technology as a whole.

In 1970 the National Academy's Committee on Science and Public Policy sponsored the establishment of COSMAT as one of a series of studies which seek to identify research priorities. Its report differs from earlier ones in part because the materials field is not yet a homogeneous and accepted discipline, like physics, chemistry, or mechanical engineering, and it does not have a professional institution covering the whole extent of the field. The committee (COSMAT) was chaired by Morris Cohen, a metallurgy professor at M.I.T., and cochaired by William Baker, then a senior research executive, now president of Bell Telephone Laboratories. Another Bell executive, Alan Chynoweth, and another academic metallurgist, Victor Radcliffe, organized the immense opinion survey, involving nearly 1 000 materials experts, which was a central feature of COSMAT's enterprise.

The Summary Report, published in 1974, incorporates a series of agreed recommendations, with COSMAT's authority behind them. The Supplementary Report, recently published, incorporates the results of the survey and also articles and statistics, sometimes mutually overlapping in their ambit, assembled by a number of specialist panels. The articles include numerous analyses, value judgments, and forecasts, and these do not have COSMAT's collective weight behind them; but having the stamp of individual style and experience, they are sometimes more instructive and compelling than the committee's collective utterances.

The report opens with an illuminating sixtytwo-page historical essay on materials and society, in which the hands of Cyril Stanley Smith and of Melvin Kranzberg are clearly discernible. The second half of the essay is devoted to an analysis of the gradual emergence of materials science and engineering (MSE) as a coherent technical field. The second chapter is an introductory survey of the role of MSE in innovation and in certain political contexts; this section draws on the work of several panels and is accordingly rather diffuse. (The Fourth Law of Thermodynamics, not widely known, lays it down that entropy is proportional to the number of authors.) It goes on to assess the manpower, both engineers and scientists, active in the field. This is very hard to do, because of ambiguities as to who belongs to the MSE field; an opinion survey of experts was used to define an "MSE score," and those categories which scored over 45 were included. The U.S. National (sample) Register of Engineers of 1969 and a similar (sample) Register of Scientists, dated 1968, were then interrogated and scaled up for the entire population, and a total professional full-time equivalent MSE population of 315 000 was thus estimated for the United States. The findings of the COSMAT Report thus affect the work of a very large number of people.

The third chapter is of quite particular interest, especially to a reviewer who is himself concerned with research. It is concerned with the multidisciplinary nature of MSE. For a start, this feature is illustrated by ten technological case histories (coated razor blades, textured materials, integrated circuits, etc.), several of them recounted in considerable depth. There has been far too little of this in the past: Guy Suits and his staff at G.E. R & D Center pioneered this historical form, and apart from the SAPPHO Project at Sussex University (comparison of successful and unsuccessful innovations, summarized in chapter 8 of the Report) there have been no detailed case-history studies in MSE. Surely an approach that is good enough for Harvard Business School should commend itself to students of other aspects of industrial practice!

Chapter 3 continues with an essay on various factors, scientific and organizational, that favor multidisciplinary "coupling." One of these is the "Member-of-the-Club" Principle: if an organization wishes to be plugged into the international information-flow circuit, it must itself inject sufficient original information into that circuit. It is not so much a question of whether all or most of that information is of immediate relevance to the organization's needs: it constitutes the club subscription. Many of the insights of this chapter are quite general and go well beyond the confines of MSE. Apart from the multidisciplinary features of MSE activities, the chapter also emphasizes the importance of the systems approach in MSE innovation.

The second volume is concerned with national goals; and it discusses in great detail the specific fields that require research. A good many of the specific recommendations in the Summary Report stem from this section of the Supplementary Report. "National goal" is a shadowy concept, and the first pages of chapter 4, which grapple valiantly with it, leave it even deeper in the shadows. A distinction is made between "goals that have been formulated at some level of government — usually at the federal level — and are embedded in a piece of legislation, an Executive Order, or a regulation, and those that rest on a less conspicuous basis, yet partake of the nature of national goals" (p. 4–2).

"Conquest of cancer"' is in the former category, "conquest of polio" was in the latter. The concept is further muddied when we are reminded that only once, under Eisenhower, was an attempt made to formulate specific national goals; when the newly assembled "goals research staff" made their first report to the Administration, they were, it seems, promptly disbanded.

The committee has not, in my view, proved that the creature exists: I suspect that national goal is a species of the genus "Nessie." There are indeed approximations to the creature: thus a most interesting section on housing indicates in detail (e.g., table 4.25) how successive pieces of legislation have altered housing targets: in a 1949 act, "a decent home and suitable living environment for every American family" was enunciated as a target. That is good, proper, and virtuous: but is it specific enough to qualify as a goal? The section goes on to discuss, in purely technical terms, what kinds of innovation are required. If the matter had been discussed in terms of goals, then it would have been necessary to balance the resources and constraints implied by one goal against those implied by other goals. Not surprisingly, this wholly political task is not attempted!

The bulk of the policy preferences for future R & D (classified according to properties, materials, and processes) are based on an extensive survey of experts, the results of which make up the rest of the volume. The survey technique itself is most interesting (though we are not told how the respondents were chosen), and it is only a pity that the response rate was not as high as one could have wished. There is an intriguing graph on page 5–15 which plots "priority for basic research" against "degree of familiarity." Not surprisingly, people are apt to push what they know best — but this graph provides a mechanism for discounting that. Page 5–31 then lists the areas thought to be most in need of attention.

The third volume deals with the distribution of MSE R & D in various industries, types of laboratory, and institutions, and analyzes the distribution of qualified manpower. Many intriguing byways are explored: examples include the technological implications of the progressive rise of labor costs relative to material costs in the building industry (page 7–90) and the nature and origins of technical standards (mandatory, voluntary, and voluntary–consensus) (page 7–93). There is some discussion of resource constraints, technical problems of extraction and of recycling, but not as much detail as the intrinsic importance of these themes might have led one to hope for. In particular, I should very much have liked to see a proper discussion of the fiscal and legislative reforms needed to encourage recycling on a really substantial scale.

The volume goes on to an extremely interesting account of MSE education and research in American Universities. This account, partly historical and partly "state-of-the-art," is so varied as to defy any kind of summary. To pick out just a few features: the major fraction of MSE research is done in departments other than MSE departments; of these (fig. 7–36), science departments are more likely to have work of relevance to MSE than are engineering departments, though in the United States, mainline materials and metallurgy specialists are classified formally as part of the engineering fraternity. (The science/engineering tug-of-war shows up at intervals throughout this chapter.) Small departments are less likely to have a high student/staff ratio than big ones, and graduate students make up a fraction of the total which, to an English observer, never ceases to be amazing. Even so, departmental block grants typically work out at $ 75 000–$ 150 000 per graduate student!

A large part of chapter 7 deals specifically with the large block-funded laboratories (e.g., those funded by the Advanced Research Project Agency, ARPA) and analyzes their successes and failures, strengths and weaknesses in great and informative detail. (One particularly unexpected finding is that most major innovations in the polymer field have not come from ARPA laboratories, and quite generally (page 7–211] doubts are expressed concerning the feasibility of integrating polymer science properly with the rest of MSE.)

One curious contradiction arises between the praise bestowed on block-funded laboratories specifically because they release staff from writing endless proposals and from the need to convince paymasters of the appositeness of their plans, and the guarded approval expressed, in the last volume (page 8–50), of the Rothschild dogma concerning the client–supplier relationship in research! But then we are specifically told that different parts of the COSMAT Report, by dif ferent authors, are not necessarily always mutually consistent.

The fourth and last volume is in many respects the most interesting of all. It is concerned with international comparisons, with respect to both research and education, discusses research institutions in consi derable detail, and finishes with an analysis in depth of the strengths and weaknesses of multinational corporations in connection with research. The volume quite defies summary, and only one or two points can be picked out here. However, much of the discussion is quite general and not specific to MSE, and deserves the widest possilble readership. To an English observer, it is gratifying to find that the accounts of the Science Research Council, of the new direction taken by the great government laboratory at Harwell, of the Rothschild–Dainton debate, and a number of other aspects of the British scene, are all accurate and make illuminating comments.

The text returns again and again to the Japanese experience. I feel sure that this is partly due to the close mutual concern between the United States and Japan, but it is also thoroughly justified by the interest and value of the subject matter, as for instance by the long statement on the philosophy of industrial R & D by the president of Sony Corporation (page 8–109). There are curious paradoxes here: for instance, many Western observers (especially the authors of the above-mentioned SAPPHO Report) have deduced that technology transfer works best "on the hoof", and that mobility of staff between institutions is vital. Yet in Japan, because of company loyalty, this does not happen. The printed word must be more effective there than it is elsewhere. The difference may also be linked to differences in national temperament:

"In the US. most scientists discover new ideas partly by contemplation and partly by informal discussion with colleagues. Ideas are then tried out by exploratory experiments, the results of which lead to further rounds of informal discussion, and so on. In Japan, such informal preliminary discussions are rarely held. It is customary for a research worker to spend a very long time in private thought before advancing his ideas to the rest of his koza [research group] at a formal research seminar where he is naturally more likely to strive for accuracy than to be speculative. Furthermore, intellectual aggressiveness is not admired in Japan and criticisms in seminars tend to be gentle." [P. 8–63]

Again, the section on the strategic aims of Japanese industrial research (page 8–90) is deeply interesting. Here, as in many parts of the entire report, the electronics industry is adduced to illustrate general points.

There is a general discussion of the relative merits of centralized planning and a pluralistic approach to research direction, in connecion with the USSR (page 8–44). The lessons to be learnt by the United States from other countries, especially Japan, USSR, Germany, and the United Kingdom, are examined with care. It is at least a comfort to discover that the criticisms of the alleged divorce between university research and industry, which are hardly perennials in Britain and are currently being voiced in the House of Commons, are widespread in other countries too, and particularly so in Japan, where industrial innovation is so successful. I hope the chairman of the Select Committee on Science and Technology at Westminster will find time to read volume 4 of the COSMAT Report! Volume 4, taken together with the beginning and end of volume 1 and parts of volume 3, make very helpful reading for a wide range of people responsible for R & D decisions, both inside and outside the field of materials science and engineering. We are all in debt to Morris Cohen, William Baker, and their numerous energetic collaborators.

87

Disorder in Crystals

Nature, **280**, 343 (1979)

Disorder in crystals. By N. G. Parsonage and L. A. K. Staveley. Pp. 926. Oxford: Oxford University Press/Clarendon Press. 1979. £ 28.

In the hierarchy of matter — gas, liquid, crystal — crystals signify order, pattern; they represent classical discipline as against romantic anarchy. "Damn braces, bless relaxes!" cried that arch–romantic, Blake. The authors of this impressive text might have taken Blake as their mentor; their crystals are for ever relaxing from a state of Roman order into one of Italian disarray; same place, changed ways.

The book is about the various ways in which crystals can lose perfect regularity without ceasing to be crystals: it is a study of local variability superimposed on a basic periodicity. The authors in their Introduction (which neatly summarises the argument of the book) distinguish three types of disorder: disorder of position, orientational disorder and magnetic disorder. These might respectively be exemplified by Cu_3Au, where copper and gold atoms are apt to trespass on each other's appointed lattice sites; NH_4Cl, in which a triad axis of the $(NH_4)^+$ ion can point in different defined directions and the ion itself can execute a click–stop rotation about a triad; and $CsCoCl_3.2H_2O$, containing one-dimensionally ordered antiferromagnetically coupled chains of tilted 'Ising spins'. Any chemist, physicist or metallurgist, however expert he may be in one or other of these very broad categories of disorder, will discover a great variety of parallels to and variants of the particular kind of behaviour he is familiar with. The book is a remarkable achievement of synthesis.

The text begins with a detailed treatment of the thermodynamic and statistical mechanical approaches to phase transformations and degrees of order: it is a mark of the combined rigour and flexibility of the authors that they do not agonise about the propriety of classifying order/disorder transformations as phase transformations. The Ising lattice theory in various dimensions receives a full historical survey, as do the various thermodynamic approaches to the order of transformation; the concept of the λ transformation is frequently cited throughout the book. Even such a technical subtlety as the bond percolation problem (if water be imagined capable of percolating along one type of bond only in a crystal, what proportion of the atoms would be wetted if the crystal were bathed in water?) gets due attention.

There is next a useful chapter on experimental techniques, such as thermochemical measurements, diffraction, NMR, IR and Raman spectroscopy, dielectric permittivity, dispersion and loss measurements, and others. The aim is to explain what kinds of information the various methods can provide, and why; no attempts are made to describe hardware or practical difficulties.

The authors then settle down to particular materials, starting with positional order in alloys. As a metallurgist, the reviewer can confirm that this chapter is reasonably up to date and quite clear in classifying and exemplifying the types of positional order. No attention is devoted here, and little elsewhere in the book, to the kinetics of order-disorder transformations, which is a major topic of current metallurgical concern; and microstructure — for instance, the curious structures result- ing from spinodal decomposition — is not considered. Again, some important concepts are not explained but simply used: the notion of 'antiphase domain' is an example. This omission of definitions happens at intervals in both the theoretical and experimental parts of the book but, given its ambitious scope, this could hardly have been avoided.

The next five chapters (432 pages) constitute the heart of the book. The first, dealing with positional disorder in inorganic compounds, covers in particular the currently important topic of superionic conductors such as an allotrope of AgI, and beta–alumina. (The authors say of Agl that "this singular crystal has often been described as consisting of fluid silver ions in a solid lattice of iodide ions, and even as a 'missing link' between the solid and liquid states" .)

A long and illuminating chapter follows on orientational disorder in salts, pride of place going to ammonium salts and including a treatment of ferroelectric salts. This is followed by a clear disssection of the exceedingly involved polymorphs of ice, most of them created by high pressure, and the various descendants of Pauling's statistical mechanics of hydrogen bonds in ice are described.

Two chapters follow on molecular solids, with special emphasis on compounds (such as perfluorocyclohexane) which contain both fixed and labile constituents. Such compounds, called 'rotator phases' or 'plastic crystals', in which some ions exhibit extreme orientational disorder, have remarkable properties such as extreme softness and very rapid diffusion, at appropriate temperatures. Hydrogen and deuterium, various linear and non-linear molecules, and many other groups are analysed, and the chapter finishes with a valiant attempt to systematise the statics and dynamics of these variegated molecular groupings. The concept of a 'glassy crystal' (in which disorder of ions on a sublattice is frozen in on cooling) is introduced. Long-chain monomer molecules are treated quite fully, but degrees of order in polymers are not discussed at all. A special feature in these chapters is the close attention paid throughout to the implications of thermal analysis and, in particular, of the entropy changes associated with order transformations.

A further chapter discusses clathrates (compounds such as β–quinol with large near-spherical structural voids that can be partially filled with 'guest compounds', which behave as independent non-interacting parasites on the host), channel compounds of urea and thiourea, and the metallic intercalates of graphite studied in recent years. An example of what can be deduced from entropy measurements is the authors' conjecture that cyclooctane molecules caught in thiourea channels undergo conformational changes, limited by the channel geometry, at sharply defined temperatures. It would be intriguing to know how polymer chains would behave if constrained within such a host.

The final chapter deals with inorganic magnetic compounds, first in quantum-mechanical generality and then intaxionomic and specific detail. The reader is taken through the luxuriancc of phenomena such as ferromagnetism, antiferro magnetism, spin waves, metamagnetism, helical spins, and their hybrids.

After this plethora, it is only proper to outline what the authors have omitted. As already mentioned, polymers are excluded. There is no reference to liquid crystals, that other missing link, which is a pity since, in terms of statistical mechanics, they show a temperature dependence closely similar tothat of the Bragg–Williams theory of positional order in alloys or the standard theory of corresponding ferromagnetic states. These omissions are not explicit justified; other omissions, such as treatment of the structural implications of non-stoichiometry, or ordering of point, line or planar defects, are made explicit inthe Introduction, on the grounds that they have been well covered elswhere. .

One can quibble at the margins about omissions and undefined concepts, but thefinal impression left is of a *tour de force* , a synthesis which has no competitor. The book deserves the close attention of adventurous inorganic and solid-state chemists, solid-state physicists (especially theoreticians) and physical metallurgists. The printing is by facsimile reproduction of typescript, but is quite clear, and in view of its size and scope the book is not overpriced.

88

Eumorphous Amorphs

Nature, **314**, 26 (1985)

Physics of Amorphous Materials.. By S.R. Elliott. Pp. 386. London: Longman. 1984. £25, $ 60. [A second edition has since appeared.]

One of the joys of contemplating amorphous materials is their affinity with paradox. *Amorphous*, etymologically, signifies *shapeless, without form;* yet the book under review has an 80-page chapter on structure. One of the key theorems general to all amorphous solids is the Kauzmann paradox, which tells us that for all glasses (a concept not quite coterminous with amorphous solid) there is a temperature above 0 K at which the entropy must decrease below that of a crystal of the same composition; Dr Elliott offers a careful discussion of the implications of this paradox, which belongs to the category of the impossible things that Lewis Carroll's White Queen liked to believe before breakfast.

Again, Dr Elliott repeatedly insists that an amorphous material has no reciprocal lattice, and this poses enjoyable obstacles to theorists who seek to understand, in particular, electron transport in these materials. But even this "reciprocity failure" may not be a universal attribute of amorphous materials, as witness the publication, a few weeks ago, by Shechtman and others, of evidence of a non-periodic metallic phase (a *quasicrystal*), neither glass nor crystal, which yet gives a sharp diffraction pattern... of fivefold symmetry!

There is a plethora of textbooks and monographs on glasses, but until recently the word has tacitly connoted oxide glasses. A leading exemplar of this large category is Glass Science by R.H. Doremus, published in 1973, which is entirely about oxide glasses, an excellent book, subject to that limitation. J. Zarzycky's *Les Verres et l'Etat Vitreux* (1982) takes this classic approach to glass science as far as it will go. [Note: An updated English version of this book has recently – 1991 – been published by Cambridge University Press under the title *Glasses and the Vitreous State.*] Lately, a flood of literature, but no textbooks, about metallic glasses has appeared. [Note: In 1984, T.R. Anantharaman published *Metallic Glasses: Production, Properties and Applications,* Switzerland: Trans Tech Publications.] Other textbooks again, notably Mott and Davis's celebrated *Electronic Processes in Non-Crystalline Materials,* have covered the theoretical physics of conduction in chalcogenide glasses.

Elliott approaches the field in a new way. He treats amorphous materials quite generally in terms of preparation strategies, ease of glass formation, the glass transition and its statistical mechanics and thermodynamics and, in particular, structure. That chapter has the clearest exposition of diffuse diffraction from glasses and of EXAFS that I have seen. Defects are also discussed; of course, a genuinely formless material could not possibly have defects, since the existence of defects implies the existence of a non-defective structure! The author pursues a resolutely quantitative approach but combines it with frequent pauses to see where he has got to in terms of a physical model. The reader is thus helped to keep track of the argument. Problems at the end of each chapter, linked always to the relevant mathematics, will help the conscientious reader to cement his understanding.

The heart of the book is a more specialized theoretical treatment of spectroscopy and electron transport, mostly with reference to semiconducting (chalcogenide) glasses but also to metallic glasses. A number of difficult topics are painstakingly set out; I have never seen such a clear discussion of the difficult concept of two-level systems. However, as the author himself specifies, the reader must be competent in solid-state physics somewhat beyond the level of Kittel's *Introduction to Solid StatePhysics,* and this distinctly narrows the readership.

This fine book has at present no effective rival. Although — inevitably — it barely touches on some aspects of some types of amorphous solid, it is yet an excellent basis for an advanced course on the amorphous state for physicists.

89

Stuff and Sense

Nature, **325**, 768 (1987)

Advancing Materials Research . Edited by Peter A. Psaras and H. Dale Langford. PP. 391. Washington, D.C: National Academy Press. $ 47.50. In Britain distributed by Wiley, about £ 45.

In October 1985, a conference was held in Washington, D.C., to mark the twenty-fifth anniversary of the creation of the government-financed Materials Research Laboratories (MRLS). Initially there were three of them — at Cornell, Northwestern and Pennsylvania Universities — but many more were set up later and over the years they have collectively exerted an enormous influence on materials science research in the United States. This book is a record of the meeting.

The opening papers are by William Baker, Robert Sproull and Lyle Schwartz, two of whom were midwives to the birth. They map out the individual pressures and institutional reponses which led to the creation of the MRLs; describe their mid-stream transfer from the Advanced Research Projects Agency (ARPA) of the Defense Department to the National Science Foundation (NSF), and the strategic changes which this entailed; and finally attempt a judgement as to the success, past and present, of the whole enterprise. The interdisciplinary 'thrust groups' (or Materials Research Groups, MRGs) favoured by the NSF to replace, progressively, the more substantial MRLs of earlier years, have focused on certain topics such as 'organic metals', ultralow temperatures, intercalation compounds and a whole range of novel materials, which are surveyed by Lyle Schwartz: he claims that many of these activities were made possible by the cooperative organization pioneered by ARPA and refined by the NSF.

Schwartz's bird's-eye view of the thrust groups' activities leads on to eleven magisterial surveys of particular fields; here, very effective use is made of limited space to convey the flavour of recent work in metallurgy, condensed-matter physics in relation to materials, ceramics, organic polymers, materials synthesis and processing, chemistry as applied to materials science, and several others. Several of these overviews are extremely impressive, marrying broad coverage with an emphasis on the most recent developments that have especial potential for scientific growth. Ductile ordered alloys, heavy–electron compounds, reptation of long–chain polymers and ceramic packaging of micro-circuits are just a few of the topics discussed. John Cahn and Denis Gratias contribute yet another survey of quasicrystals (a mature subject, yet barely three years old). In this, the main section of the book, one has the sense that for once achievement has been as impressive as promise — the entire 270 pages can be recommended to labourers in the materials science vineyard who seek surveys of the interdisciplinary varieties grafted on to the solid, disease–resistant rootstock of the central disciplines.

A final short section, by a number of authors from many industries and universities, discusses some of the hardy perennials of science policy. Some but by no means all of this section is restricted to purely American issues. Small science versus large science and the grave problems of getting equipment and financing support staff in the MRLs and MRGs are debated here, as is the expected role of particular families of materials and processes in particular industries (near-net-shape manufacture, ceramics in heat engines, surface modification, molecular-beam epitaxy and so on). Several contributors make points which would risk provoking cynical shrugs in Britain today: Albert Clogston, late of Bell Laboratories, insists on the crucial importance of *basic* technology in research and development laboratories, while Praveen Chaudhari, a senior research manager at IBM, declares that "an important point which cannot be taken for granted or emphasized enough is that the research enterprise of the nation requires an infra-structure that nurtures general science, or science that cannot be identified at present with any particular area of application". This comes from an industrialist, not from an academic!

Earlier in the book, Sproull emphasizes how the MRLs were based on the post-war recognition that Government contracts and grants "should provide much more scope for the contractor's imagination and discovery and more harvesting by informal agency–contractor interactions rather than by fulfilling specifications", and that the Office of Naval Research (one of the progenitors of the MRLs) favoured "task statements in general terms, with maximum opportunity for creation and discovery". In spite of the temporary setback of the Mansfield Amendment, the United States has by and large been able to keep these ideals. Certainly, it seems that Senator Mansfield has had a less baleful influence than Lord Rothschild in Britain.

For non-American readers, there are some *longueurs* in the book because it concentrates exclusively on the United States and its achievements. But the diet is by no means an unvaried one of self-congratulation. DiSalvo bewails the neglect of the creation (synthesis) of novel bulk materials and compares it with the French penchant, orchestrated by the Centre National de la Recherche Scientifique, for combining synthesis of novel materials with immediate assessment of properties, often in the same laboratory. This is one of several places in the book where the need is pressed for materials scientists to give more attention to solid–state chemistry.

This volume represents a landmark in the literature of materials science and engineering. It deserves a wide readership, not only among materials practitioners but also among specialists in science policy.

90

A Particular Way With Words

Nature, **330**, 292 (1987)

The Art of Scientific Writing: From the Student Report to Professional Publications in Chemistry and Related Fields.. By Hans F. Ebel, Claus Bliefert and William E. Russey. Pp. 493. Weinheim: VCH. Hbk, DM 98, $ 59.95, £ 39.25. Pbk, DM 48, $ 24. 95, £ 19.25.

People who draft patents are apt to refer to the 'state of the art', by which they mean current technique, technology, design, disposition, proportion, processing, assembly and the like. The book under review covers the state of the art of scientific communication, and going beyond that it also surveys the strategy and tactics of English style as practised by skilled (and unskilled) scientific communicators.

Books of this kind fall into two broad categories: reference manuals (such as the *Chicago Manual of Style* or *Copy–editing — The Cambridge Handbook*, by Judith Butcher) and discursive guides to good usage (such as *The Chemist's English*, by Robert Schoenfeld). Although Ebel and his colleagues have contrived to steer a median course between these extremes, they do aim more towards systematic exposition and exhaustive instruction. They also seek to explain why as well as show how, and in these stated objectives they have succeeded.

The authors have a passion, not only for clarity and economy of style, but also for precision and consistency. Their attitude to both scientific writing and editing is reminiscent of the words of William Blake: "He who would do good to another must do it in Minute Particulars. General Good is the plea of the scoundrel, hypocrite and flatterer; for Art and Science cannot exist but in minutely organized Particulars". (I wonder...is that last full point in the correct place relative to the quotes?)

The book treats, essentially, three broad themes: how to organize a report, dissertation, paper or book and how to achieve maximum clarity in the enterprise (the detailed advice on dissertations will be especially useful to anxious graduate students, who should find the paperback price accessible); the practicalities of drafting, correcting, typing or printing, drawing illustrations and chemical molecules, and proof-reading (the reasoned advocacy of computers as word-processing tools is particularly helpful); and the accepted conventions of citing the literature, using units correctly, setting out mathematical expressions, laying out tables, abbreviation of scientific terms, and chemical nomenclature. Although the authors are chemists and pay special attention to chemists' difficulties, other readers merely need to do a little judicious skipping. In addition, there are 100 pages of appendices, mostly factual but also including a concise discussion of good English usage.

The authors certainly practise what they preach: I found only a single misprint and one grammatical solecism ("x is comprised of y").

On just one aspect of style the authors, uncharacteristically, blow both hot and cold: this is the vexed issue of the active versus the passive voice. While themselves preferring and occasionally using the active voice, they claim that their task is to describe the world as they find it and that most of us prefer impersonal (and thus passive) constructions when describing our work.. "The mixture was distilled" I accept. But I think that the phrase "It is thought that" is an abomination. We all of us, in writing about science, describe what was done and say what we think, and why. Putting opinions in the passive voice seeks to impose a semblance of objectivity on something that by its nature cannot be objective. As a former editor of the *Guardian* newspaper once (nearly) wrote: "Facts are sacred (and passive), opinions are free (and active)".

91

Way out of the Waste Land

Nature, **339**, 24 (1989)

Radioactive Waste Forms for the Future.. Edited by Werner Lutze and Rodney C. Ewing. Pp. 778. Amsterdam: North-Holland. 1988. Dfl. 470, $ 247.25.

When one reads a diatribe against nuclear power, one often finds the bald assertion that no way exists of disposing of radioactive wastes: as long as nuclear reactors are allowed to exist, wastes must relentlessly accumulate, we are told, and finish up by killing us all. Those who detest and are afraid of nuclear power echo the poet and critic William Empson:

Slowly the poison the whole blood stream fills.
It is not the effort nor the failure tires.
The waste remains, the waste remains and kills.

The contributors to this book show, in exhaustive detail, why the waste remain and does not kill.

When a charge of nuclear fuel is 'spent', it is discharged into cooling ponds so that its radioactivity can diminish over a periodof a few years to a level which permits safe remote reprocessing treatment. The remaining radioactive fission products (high-level wastes) can then be separated out and concentrated in solution; many thousands of litres have been stored for some years in this form while the best long-term strategy for permanent disposal in solids was being developed. Such a strategy is now generally agreed and has been put into effect in many countries. The assertion that no such method exists is simply untrue.

The editors claim that this is the only book which systematically pulls together the enormous amount of work done over the past 30 years on the alternative ways of 'immobilizing' high-level nuclear wastes. Much of that research has been published in the 'grey literature' of final reports from various national laboratories or in conference proceedings, many of them not easily accessible.

The procedure which is now in large-scale operation in France, Belgium (in a plant built by Germany) and India, which is about to start large-scale operation in the United States, Britain and Japan, and which operates at pilot-plant scale in Canada and Italy, is *vitrification*: that is, the incorporation of the waste within blocks of borosilicate glass which are then further encapsulated in welded metallic containers, for later burial in deep tectonically inert caverns (initially in reversible disposition, later in permanently sealed sites). The principle of multiple barriers is crucial to this strategy: the glass itself, the outer container, the inert locations all have their parts to play. A number of other strategies have been the subject of much experimental attention, most of them involving various forms of ceramic vehicle instead of glass (notably the Australian Synroc and the American tailored ceramics). All of these are covered in the book, which however does not discuss the geological character of suitable burial sites; this topic has been covered elsewhere.

Ten Chapters cover the various types of immobilization, the most comprehensive of them being Lutze's 160-page survey of borosilicate glasses. There is also one Canadian contribution which discusses the case for leaving spent fuel unreprocessed. A twelfth chapter, by the editors, compares the glass option with the others, primarily with Synroc. This splendid essay forms an appropriate conclusion to the book.

The editors insisted upon a standardized chapter format, to ease critical comparison between the various options. Not only are all the obvious themes covered, such as leaching behaviour, production methods, response to radiation damage and radioactive self-heating, and physical and mechanical characteristics, but evidence is also adduced from natural analogues (such as basaltic glass and palagonites) which allows the long–term chemical stability of man–made materials to be estimated by comparison. Each contribution was scrutinized by a band of 39 reviewers from nine countries, and apparently several of the chapters were extensively revised as a consequence.

A reading of the book makes it clear why, in spite of a plethora of options, all countries which have built or are building solid immobilization plants have chosen the glass route — the amount, variety and reliability of the accumulated information about this system, including production methods, exceeds that about its competitors; and although other systems have some points of superiority (for example, Synroc's lower solubility in ground water), in other respects (for example, smallness of volume change on irradiation) glass scores unambiguously. Nevertheless, as the editors point out, the availability of a numher of "fairly thoroughly developed alternative waste forms" constitutes an "excellent situation".

This is a critical and unemotional overview of a very large body of information. It will be essential reading both for scientists working in the field and for others who wish to arrive at a fact–based judgment on the degree of danger associated with high–level wastes.

92

Materials Engineering for Nobody

Angewandte Chemie (*Materials Suppt.*), **100,** 1033 (1988)

The Technology and Applications of Engineering Materials. By M. S. Ray. xxv + 736pp. Englewood Cliffs, N. J., U.S.A.: Prentice Hall. 1987. Pbk., $ 27.95. ISBN: 0-13-902081-0.

Many years ago, a review of another book on materials opened with the words: "Of books on Shakespeare's plays,the Malaise of Modern Man, and materials science, there is no end. No blinding new insights are to be expected on any of these topics: selection, clarity and economy must be our touchstones." Since I wrote those despairing words in 1974, my shelf of such books has grown to bend increasingly under the burden of several dozens of such volumes (see the book under review, pp. 455-60, "Bending of Beams"). Any new general text on the science and technology of materials has to be of exceptional quality to succeed in a crowded market place. Martyn Ray's book, I regret to say, fails the tests of selection, clarity and economy: I cannot give it even a qualified welcome.

Texts for students of Materials Science and Engineering (MSE) set out with different objectives. At one extreme is the physics–centered, fundamental interpretative approach, setting out to explain properties in terms of atomic and crystal structure, quantum mechanics and statistical mechanics. The first such book was Cottrell's *Theoretical Structural Metallurgy* of 1948, followed some years later by Wert and Thomson's *Physics of Solids* (1964). Another, even earlier, and extremely influential physics-based book was Barrett's *Structure of Metals* (1943), which has lasted well in the form of successive editions. The same fundamental approach is possible in regard to the chemical approach to materials. The classic here is Darken and Gurry's *Physical Chemistry of Metals* of 1953. Midway between the physical and chemical approaches one finds Swalin's *Thermodynamics of Solids* (1962).

The fundamental approach to polymers was a little slower in coming and probably began with Treloar's 1958 book, *The Physics of Rub-*

ber Elasticity. Since then, the polymer scientists have also been well served with fundamentally biased texts, with a strong undertow of statistical mechanics. From the 1970s on, basic texts began to use the term 'materials science'. Notable examples include Ruoff's *An Introduction to Materials Science* of 1972 and Hornbogen's *Werkstoffe* of 1973.

At the other extreme there are texts aimed at the engineer who selects materials for incorporation in his designs, and also at the materials processing specialist who converts materials into semi-finished or finished products. These have been, if anything, even more numerous than the science–centered texts. Early ones had titles such as *Metallurgy for Engineers* (Rollason, 1939) and *Physical Metallurgy for Engineers* (Clarke and Varney, 1952). Later titles covered a broader range of materials; among the better are *Engineering Materials* by Jastrzebski (1959), The *Principles of Engineering Materials* by Barrett, Nix and Tetelman (1973), *Structure and Properties of Engineering Materials* by Harris and Bunsell (1977), as well as the comprehensive *Metals, Ceramics and Polymers* by Wyatt and Dew-Hughes (1974), which was the volume that drew forth the revie er's words cited at the outset.

A good example of a book aimed specifically at processes is Alexander and Brewer's *Manufacturing Properties of Materials* (1963). More recently still have come some splendid texts aimed directly at developing for fledgling engineers a systematic approach for selecting materials during the design process: *Engineering Materials — An Introduction to their Properties and Applications*, by Ashby and Jones (1980) is a good example.

The titles cited here are only a small selection and include some of the best and most durable. Some of them, especially in the 'engineering' group, are really broad treatments that succeed in marrying the S and E of MSE: the books by Wyatt and Dew-Hughes and by Harris and Bunsell are examples. The engineers who are to learn from these books are paid the compliment of being supposed intelligent and curious, of wanting to understand complex facts rather than to learn them by rote (and then promptly forget them). By contrast, Ray's book (the volume under review) assumes that his readers do not want to understand — or, perhaps, are incapable of understanding? I am sorry to have to add that this attitude may well stem from a frequent and unmistakable lack of understanding by thc author himself. A diet of bare uninterpreted fact, often so vague as to be unintelligible even as fact, has to substitute for insight.

A few quotations will give the flavor: "The molecular weight used to characterize a polymer may be based upon several criteria. The most popular are the viscosity average, number average and weight average molecular weights. For a molecular weight distribution these averages have different values, although they would be identical if the polymer possessed a unique molecular weight." That is all on this subject. Another quotation: "The creep process occurs because of two mechanisms; these are grain boundary sliding and dislocation movement by climbing past obsta cles." That is all there is in the book on the role of dislocations in plastic deformation... yet the cover is decorated by a large diagram of an edge dislocation! Another: "The TTT diagrams described so far are obtained by cooling a steel isothermally at a series of tcmperatures." One of the illustrations in the chapter on joining consists of a variety of screw heads!

Several features of the book, especially the very extensive lists of standard specifications and many tables of numerical values of properties, betray a measure of confusion between the roles of a textbook and a handbook (such as Smithells' *Metals Reference Book*). 166 pages are devoted to elementary school mechanics, which have no place in a book of this kind, supposedly directed at professional engineers–to–be. According to the preface, the book is in fact directed not only at these but also at technicians taking pre-degree courses. But these, also, need and deserve understanding, not just a diet of undigested and sometimes erroneous fact.

Ray also has the peculiar habit of providing a list of keywords for each chapter, apparently in the hope that the reader can check for himself whether he has learned key concepts. In a chapter on engineering design, the list of keywords includes, inter alia, "perseverance, willpower, scientific knowledge, conceptual ability". The author has demonstrated the first two but not, alas, the last two.

93

Ternary Alloys

Advanced Materials, **1**, 128 (1989)

Ternary Alloys — A Comprehensive Compendium of Evaluated Constitutional Data and Phase Diagrams. Volumes 1 and 2, *Ternary Silver Alloys*. Edited by G. Petzow and G. Effenberg. Principal reviewer: H. L. Lukas. Vol. 1, 612pp., Vol. 2, 624 pp. VCH Verlagsgesellschaft, Weinheim, Germany and VCH Publishers, New York, N.Y, U.S.A. 1988, $ 250 per volume (special arrangements for series subscribers). ISBN 3-527-26941-X and 3-527-26942-8.

Metallurgists, like any other kind of scientist, have numerous useful tools of their trade, and a few indispensable ones. Prominent among the indispensable ones are phase diagrams. They are equally necessary to researchers who seek to develop new alloys or improve existing ones, and to producers who seek to optimize the properties of standard alloys by modifications to manufacture or heat-treatment. Modern metallurgy would simply not exist without them.

The first really accurate phase diagram (Cu -- Sn) was determined by Heycock and Neville, two Cambridge chemists, and published by the Royal Society in 1904. Only 32 years later, a notable German pioneer, Max Hansen, published *Der Aufbau der Zweistoff–Legierungen*. This thick single volume reported on the phase constitution of 828 metallic systems and included 456 critically evaluated phase diagrams. After the war, three successive revisions and supplements of this work appeared and other critical compilations of binary phase diagrams were published subsequently. The key word is "critical": somebody has to make judgements between conflicting data and interpretations. To organize and institutionalize, on a worldwide basis, the heavy task of assembling information, evaluating and reviewing it, an Alloy Phase Diagram International Commission (APDIC) was set up some years ago and a journal, *Bulletin of Alloy Phase Diagrams*, instituted.

One might suppose that the availability of so many binary phase diagrams would satisfy the needs of most metallurgists, but in fact alloys become ever more complex, and many contain at least three major constituents. *Ternary* diagrams have, therefore, become increasingly important in recent years, and many have been evaluated for, and published in, the *Bulletin*. In its last issue, an editorial tells the reader that up to now, 635 binary systems have already been evaluated and published in the journal, its Indian sister journal, or in books, and over 400 more are in progress. In December 1988, the *Bulletin* informed the reader, 606 ternary systems had also been evaluated and published, and 408 had been evaluated but not yet published. As a proportion of possible ternary systems (bearing in mind that more than half of the elements are metals) this is still only a very small proportion, but it must be borne in mind that the effort involved in determining a ternary system is much greater than for a binary. (The great English metallurgist R. S. Bradley, for instance, spent over a decade of single–minded labour around the middle of this century on the determination of the Al-Fe-Ni diagram alone).

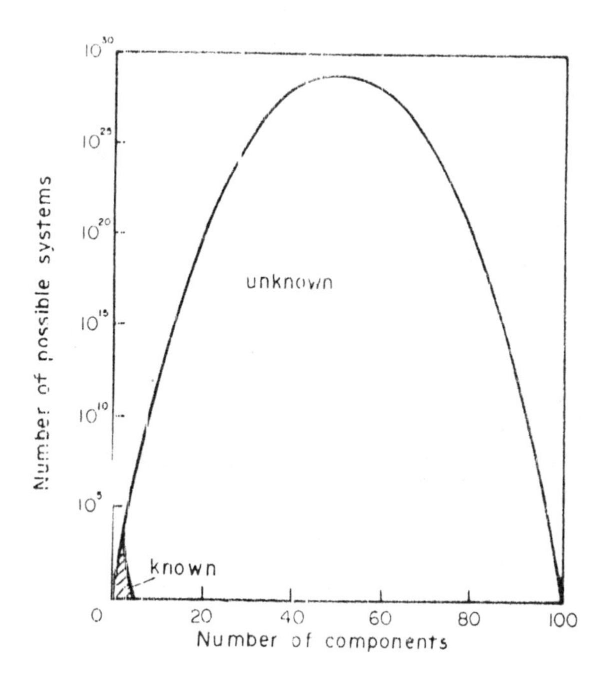

Fig. 1. Numbers of known and unknown heterogeneous phase systems. (After G. Petzow, 1986)

The ≈ 2000 binary and ternary phase diagrams now reasonably well established (many others have only been cursorily surveyed) are a truly tiny fraction of the possible total. Figure 1 (note the logarithmic ordinate scale!) was published recently by one of the editors, G. Petzow, of the volumes under review. Allowing for 100 elements (including a few transuranics) the number of possible 50–element systems exceeds by a large margin the age of the universe in seconds. Fortunately, determination of 50ary phase diagrams would run into such formidable problems of technique, evaluation and depiction that we can comfortably forget about them; even quaternary and quinary systems, which are occasionally published, are hard to determine and harder to understand when published. What really matters to practitioners is the availability of numerous reliable and conveniently available ternary systems. The new series on which the German evaluators and the publishers have now courageously embarked clearly will fulfil this need.

The two volumes under review incorporate critically evaluated data for 446 ternary systems containing silver, critically updated to January 1988 and based on over 1900 publications of experimental work, taken from the huge databas maintained at the Max-Planck–Institut für Metallforschung in Stuttgart, with German Government support. Though the calculation of phase diagrams from thermochemical data has made great strides recently, this approach is apparently not yet being used for these volumes, because of the very high standards of accuracy which are being adopted and because of the difficulty in securing adequately precise 'optimization' for the three constituent binaries of each ternary system, which is a prerequisite for ternary calculations. For many though not all of the 446 systems, ternary diagrams are included; related evaluated binary diagrams are also included where these are not readily available in one of the standard volumes of binaries already published.

Each system is the responsibility of an individual evaluator, 20 of whom contributed to these two volumes and although half of these 20 are German, many nationalities are represented. Each evaluation was then critically reviewed by one of a small band of reviewers. This has become universal practice in all assessments of ternary diagrams under the broad aegis of APDIC, as the present volumes are. The reviewers and the original evaluators, if need be, met to resolve their differences at one of the numerous editorial conferences held at a specially provided rural idyll.

The task of producing volumes like these is truly daunting, quite apart from the successive levels of evaluation and reviewing to which all the work is subject. Finding the original publications (many in exceedingly obscure periodicals); assessing their reliability; turning the diagrams into standard format; finding and assembling crystallographic information about the phases (structure types and lattice parameters are cited); creating isothermal and 'vertical' sections; in particular, drawing up the involved 'reaction schemes' for the more complex ternary systems which have become *de rigueur* for serious depictions of ternary diagrams, to render them more intelligible; computer–based creation of final drawings and checking of the final printed form... all this has called for a very large team of participants both in Stuttgart and elsewhere, and at the publishers a very long term commitment to a colossal task, as well as generous Government funding without which, the (admittedly) high price of the volumes would have gone right through the roof.

The stated objective of the series is to have so much information not, just cited but presented in the volumes, that most users will never need to get to the original literature. In this reviewer's opinion, that objective has been achieved.

The next set of volumes, three in number, on ternaries containing aluminum, is at an advanced state of preparation and is due to be published this year. [Note: In the event, publication of these volumes, four rather than three, has been spread over the years 1990–1992.] It should be noted that as the series progresses, the task will become gradually less onerous. This can be appreciated by noting that the two 'Ag' volumes incorporate 28 Ag-Al-X systems, and none of these 28 need be reevaluated for the 'Al' volumes, though they will be printed again there. This 'effort reduction effect' will rapidly become more significant as successive vol umes are published.

There have been other compilations of ternary phase diagrams in book form. First, I should point out a book entitled *Phase Diagrams for Ceramists* , published by theAmerican Ceramic Society, originally in 1964. This contains information on many binary, ternary, quaternary and even a few quinary systems of ceramics. The ceramists are entirely separate from the APDIC organization. (There are no compilations of phase diagrams of polymers, and indeed very few such diagrams exist. The concept of phase equilibria is only making slow headway in the polymer world, though both theorists and experimentalists are now working extensively on the concept of miscibility and separation of phases in 'polymer blends').

India has played a large part in the international metallurgical effort, and many evaluations have originated there; Raghavan in Delhi has published three volumes on Fe ternaries (the latest has just appeared) with one more to come. Even so, the four Indian volumes only scratch the surface of the enormous range of Fe ternaries...i.e., steels! In America, a Cu-O-X volume is expected to appear soon (again, clearly, a small subset of Cu ternaries) while in Britain, as a result of work initiated years ago by that prince among evaluators, the late Geoffrey Raynor, a volume of Au ternaries is expected to bepublished soon by the Institute of Metals. The editor of this is Alan Prince, who is now the doyen of British evaluators and has played a substantial part in the two Ag volumes under review here. When the Stuttgart team gets to Au, which should not be long in view of its position in the alphabet, there will undoubtedly be arrangements to make full use of the British work, to avoid needless duplication, but presumably the information in the British volume will all be included in the VCH series, which is intended to be as comprehensive as is humanly possible.

It is entirely appropriate that the massive international effort which has created this first fruit, the silver volumes, should be centred in Stuttgart. Many of the early phase diagrams were determined by the German pioneer, Gustav Tammann. Germany has long had an indigenous tradition of high-class metallography, including a training programme for metallographers which has no equal; metallography (and the ability to judge critically the results of metallographic investigations) is an essential constituent of phase diagram determination and evaluation. Max Hansen, the father of phase diagram collection and evaluation, worked at the Stuttgart Institute, and as the editors of the silver volume indicate in their preface, they are seeking to create a sort of 'ternary Hansen'. In this ambitious task, to judge from its first fruit, they have triumphantly succeeded, and the publishers also deserve every credit for their courage and long-term commitment in bringing this venture to the public.

94

The Breakthrough

MRS Bulletin, **14**/1, 75 (1989)

The Breakthrough . By Robert M. Hazen. 271 pp. New York: Summit Books. July 6, 1988. $ 18.95. (Published in the U.K. by Unwin Hyman on 22 September 1988).

The history of science, like the history of political events and war, has become steadily more variegated; the newest format is instant history. The book under review is a distinguished addition to this subcategory.

Roger H. Stuewer's book, *The Compton Effect* (1975), is a classical book about the origins of a modern scientific phenomenon — an eminently respectable text with an extensive bibliography of books, papers, and letters. Perhaps more suspect to traditionalists is Daniel J. Kevles's *The Physicists* (1971), a study of the growth of a scientific profession in one country. The source of possible suspicion is Kevles's extensive use of newspaper and magazine sources. Kevles did not go for interviews, however, unlike T.R. Reid, author of *The Chip* (1984). A journalist, Reid wrote his fine study of the invention and apotheosis of the integrated circuit on the basis of interviews, as well as books and published papers, all of which he cites.

The extreme, till now, of this less conventional approach to sources was James D. Watson's *The Double Helix.* In 1968 this 'instant' history, based on memory and notebooks...with no bibliography...and with its candid delight in the fact that scientists share the foibles and frailties of the rest of humanity, proved a shock. The editor of *Nature* could not persuade any biologist to review it and the task fell to a professor of comparative literature! Perhaps this is why the *MRS Bulletin* 's editor thought it politic to approach someone who has never worked on superconductors to review Robert Hazen's gripping book, in some respects closely akin to Watson's book.

MRS Bulletin readers will scarcely need to be reminded of the scientific stampede unleashed when Bednorz and Müller's cautious, low-profile paper of September 1986 became widely known on the occasion of the MRS [Note: MRS denotes the (U.S.) Materials Research Society] superconductivity symposium of December 1986. During the intervening weeks, only a few scientists took in the full significance of what Bednorz and Müller had shown — that a ceramic (and ceramics are usually insulators) had driven up the superconducting transition from ~23 K (for the best alloy) to over 30 K. Paul Chu of Houston and his team quickly improved this figure and their breakneck labours led to the Y-Ba-Cu-O superconductor, with $T \approx 90$ K.

It then became extremely urgent, first to identify the composition of the superconducting phase in the polyphase ceramic, and second to determine its crystal structure. Chu was convinced that the crystal structure would be the key to improving and, eventually, to understanding high temperature superconductivity. Chu experienced a personality clash with his local x-ray diffraction expert and so turned to David Mao, a member of the informal 'mafia' of Chinese-American scientists (the thumb-nail sketch of which is one of the book's many minor felicities). Mao, at the Geophysical Laboratory in Washington, at once gave the task of structure determination to his colleague Robert Hazen, who in turn brought in further co-workers.

Formally, the book covers the month from February 20, 1987, when Hazen accepted the task, to March 18, the date of the American Physical Society meeting which homed in on the $YBa_2Cu_3O_x$ ("1-2-3") superconducting phase. *De facto,* the book looks at the antecedents of the fevered month and glances sideways at simultaneous happenings elsewhere in the world. The main emphasis, however, is on Hazen's frenetic, nightmarish, but wildly exciting period of unremitting labour.

Aristotle ruled that successful drama requires the writer to observe three unities — action, time, and place. Hazen splendidly achieves the first two, but he could not achieve the third, unity of place. Most of the participants, especially Chu, travelled extensively during that anxious month, speeding from laboratory to laboratory, from laboratory to meeting, and occasionally even home.

Hazen's book is splendidly constructed. It helps, no doubt, that he is a multiple professional: a professional X-ray crystallographer, a professional trumpeter (an apt instrument), and a semi-professional writer. He sets the scene, explains the essentials of superconductivity and its associated measurements, and offers thumbnail sketches of numerous protagonists of the 1-2-3 hunt. He describes how the 1-2-3 phase was identified and isolated, and the process of structure determination, with a canny use of metaphor (the difficulty of interpreting an x-ray powder pattern of two mixed phases is likened to the problem of exploiting two superposed fingerprints).

All this leads to the dramatic climax of the New York meeting of the American Physical Society, with its four rival structure accounts (and the steps that led to reconciliation of the apparent disagreements). The meeting is described in a splendid piece of atmospheric writing, as cliff-hanging as a good whodunit. The science is readily accessible, at a superficial level at any rate, to any intelligent reader, scientist or not.

The book contains not a single citation, but differs crucially from *The Double Helix* in that Hazen took pains to check the accuracy of his version of the involved sequence of events with many of the principal protagonists — quite an achievement in a book published only 15 months and 18 days after the New York meeting. (This is why I have thought it right to cite, in the heading, the day and month as well as the year of publication!)

Since published papers always tidy up the chaotic sequence of actual research, Hazen's is presumably the only approach that can hope to present what actually happens, step by step, in the heat of the moment. Incidentally, while the book recounts some apparently disreputable episodes of scientific espionage (which proved false alarms), it has nothing to bring a blush to the most modest cheek

The main feature of that central month was a race to be the first to identify what proved to be 1-2-3, and to be first with its crystal structure. The unspoken assumption is that fame, fortune, and a Nobel Prize all depend on submitting a paper to *Physical Review Letters* a day or two before any rival. It is quite an assumption!

Recent history might have reminded the protagonists that in the 1984 fever-pitch excitement over the new permanent magnet superphase, $Nd_2Fe_{14}B$, four teams independently determined its crystal structure and published it almost simultaneously. Now, when this compound is discussed in the literature, all four papers are co-cited; no one inquires which was submitted the earliest.

The frenzy of the race for 1-2-3 was perhaps out of proportion, but perhaps essential to provide the free energy for the achievements of early 1987. Some did manage to keep a sense of proportion. Despite hordes of physicists clamouring for admission, the hotel management refused to release the largest ballroom to the APS meeting because it had been booked for a wedding reception!

MRS members, almost by definition, are devoted to research. Regardless of their immediate scientific concerns, this book is confidently recommended to all of them.

95

Image Analysis

Advanced Materials, **2**, 111 (1990)

Quantitative Image Analysis of Microstructures.. Edited by H. E. Exner and H. P. Hougardy. 235 pp. Oberursel, Germany: DGM Informationsgesellschaft Verlag. 1988. Hbk., DM 95, $ 68. ISBN 3-88355-132-5 (English Edition)

I begin this review by referring to two figures, both borrowed from the book under review. The subject-matter of the book is image analysis, so where better to start than with two images, each of which represents the analysis of many images?

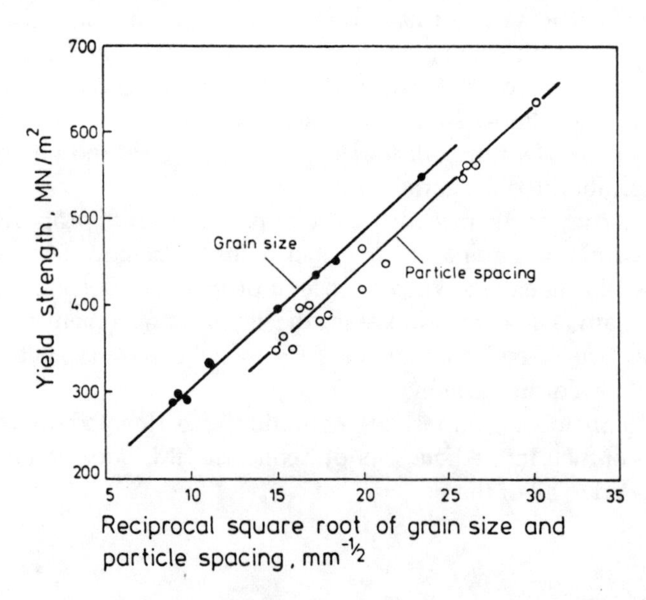

Figure 1. Yield strength in relation to grain size or particle spacing.

The first figure concerns a very common issue in physical metallurgy...how does strength relate to the average scale of a microstructural feature? The graph consists of two straight lines, one referring to the size of crystal grains in a low-carbon steel, the other to the spacing of cementite (Fe_3C) particles in a higher-carbon steel. The manner of plotting these variables — using their reciprocal square roots — has been common with measurements of this sort for the past 35 years, and arises from an analysis of the way in which dislocations pile up against obstacles such as grain boundaries. The size of such pile–ups in turn determines the strength. This kind of matching of theory and observation has led to important advances in alloy design.

At first sight, nothing could be simpler than to measure the size of crystal grains; after all, they are straightforward to etch and thus are easy to see under an optical microscope. Then, perhaps, one recalls that crystal grains in an alloy have a distribution of sizes and may or may not be elongated in one direction. If (as is customary) one estimates the grain size by drawing a randomly positioned line across a micrograph and counting the number of times a given length of line intersects grain boundaries, one must then allow for the fact that the cross-section from which the micrograph was made does not cut most grains across their middles and thus makes them look smaller than they really are; that the line does not normally cut a sectioned grain of irregular shape across its maximum diameter; that in deriving an average grain diameter, one must allow for the possibility of a directional dependence of this average and that (when one has corrected for the other sources of possible confusion just cited) there are different ways of defining an average for a distribution of grain sizes, just as one can quote a mass average or a number average molecular weight for a high polymer. Again, the result will be the more reliable, the longer the length of the probing line; just how much more reliable? As for the spacing of particles, this can be defined in terms of average separation of nearest neighbors or of a mean free path, which is much greater. — Exner and Hougardy's book deals with these kinds of problem, a complex mix of definitions, measurement techniques, interpretation, statistical considerations and instrumental design.

The second figure shows measurements of the scale of dendrite configurations (patterns akin to snow crystals) in slowly frozen superalloy specimens. This kind of linear plot can then be extrapolated and used to estimate cooling rates of ultrarapidly quenched powders of the same alloy; there are in fact no other accurate ways of estimating cooling rates of 10^6-- 10^8°/s. Here, then, image analysis is applied as an indirect means of measuring a quantity of practical interest in materials processing.

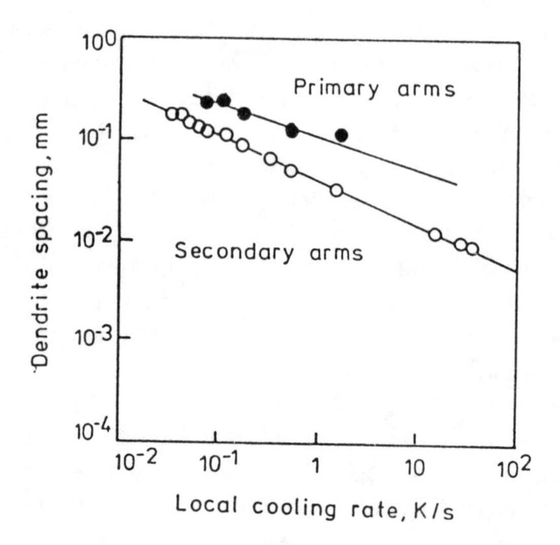

Figure 2. Dendrite arm spacings in a superalloy as a function of solidification rate.

Exner and Hougardy's book is based on a practical course given repeatedly at the Max-Planck-Institut für Metallforschung in Stuttgart. Not only uses of image analysis in metallography and ceramography (the central concerns of the Stuttgart Institute) but also in the earth sciences and in the life sciences, were treated, and five research establish- ments apparently helped to put the course together. Exner (of Stuttgart) and Hougardy (of the sister establishment in Düsseldorf which is devoted to steels) are described as *editors* on the title page but as *authors* in the preface. Since no other contributors are named, I think we must regard them as authors! The book treats the mathematical and statistical fundamentals (including the subtle but virtually incomprehensible French approach called *mathematical morphology*); the generalities of obtaining reliable information about circumstances in three dimensions from two-dimensional sections and projections; a review of measurement techniques, including charts, graticules, etc.; and a very comprehensive survey of available instrumentation, ranging from the nearly manual to the totally automated, and a critical discussion of the bases on which a choice should be made. Even the metallographic methods to be used to avoid accidentally falsifying such experimental variables as volume fraction of a second phase are critically reviewed. Applications, in metallography only, are exemplified in depth.

The book is copiously illustrated and there is an extensive bibliography. The text is clearly reproduced from camera–ready copy and is fairly easy to read, but by no means as easily as typeset text. The current cost-cutting practice of combining camera–ready copy with size reduction is getting to be a vice in the world of science publishing, especially when it is combined with a price which is far from modest in relation to the number of pages. Could not the country which invented movable type revert to using it more, even if in America this seems to be a lost cause?

This little item of carping to one side, the book under review is a splendidly authoritative, clear and useful survey of a field which more materials scientists could with advantage learn to exploit. Quantitative image analysis, alias quatitative metallography, alias stereology, is a wonderful way to convert hand-waving into finger–pointing.

96

COSMAT Revisited

Advanced Materials, **2**, 206 (1990)

Materials Science and Engineering for the 1990s — Maintaining Competitiveness in the Age of Materials.. By the Committee on Materials Science and Engineering of the National Research Council (Praveen Chaudhari and Merton Flemings, chairmen). 294 pp. Washington, D.C.: National Academy Press. 1989. Hbk., $ 39.95. ISBN 0-309-039238-2.

This is a fascinating book... but the fascination is directed more at politicians, captains of industry and those who dispense government funds than at materials scientists and engineers. It is also a book which has enjoyed a substantial build-up of publicity through published interviews with the joint chairmen of the authorial Committee. Disraeli once remarked that "an author who speaks about his own book is almost as bad as a mother who talks about her own children", but since the volume under review has no identifiable authors, or even editors, and has indeed been written by a Commitee, Disraeli would be hard put to it to censure these premonitory interviews! While it is often said that a camel is a horse designed by a Committee, it is also true that the King James Version of the English Bible, the finest flower of English prose, was written (or at least approved) by a committee... so everything depends on what kind of committee has done the writing (or approving). The Committee of Materials Science and Engineering of the National Research Council, aided and abetted by a plethora of supporting panels, subpanels, steering Committees, boards, commissions and individuals, has done a pretty sound job of preparing a polished end product. This volume has been very well synthesized and processed, but one would rather like to know what skilled writer prepared the final draft so that one could pat him on the back.

What is here on offer is a successor to the COSMAT Report of 1975; this was a survey entitled *Materials and Man's Needs* (see article 86) and sought to explore what Materials Science and Engineering (MSE) is, where its practitioners come from, how the field is practised in various parts of the world and how it relates to National Goals. The COSMAT Report was much longer and detailed than the present volume, it was written by a single small committee and it sought to persuade the scientific community that MSE really exists. Those who practice this field (all except a few diehard metallurgists) no longer need convincing of this, but it seems that the politicians do. In fact, the 1989 volume was sparked off by an initiative by a Congressman, Don Fuqua, who in 1984 wrote a letter to the presidents of the two National Academies inWashington requesting a "comprehensive materials research and technology assessment for the next decade". The federal agencies which support MSE research cried "amen" to this, and the slow progress from council to committee to board, from board to committee to council was set in train, until finally the Committee which actually prepared the book was constituted. As we see, the result appeared just before the last decade was out, thereby meeting Mr. Fuqua's objective. The joint chairmen are, respectively, Vice-President for Research of IBM and chairman of the MSE Department at MIT, and they have ensured that industry and academia have played balanced roles in putting together the text.

The first 18 pages of the book consist of a Summary (what captains of industry call an Executive Summary), a set of conclusions and a longer set of Recommendations. The conclusions are grouped under the headings Role of Materials in Industry, [Research] Opportunities in MSE, Emerging Unity and Coherence of the Elements of MSE, Instrumentation and Modeling, Education, Infrastructure and Modes of Research, Federal Support for MSE and MSE in Selected Countries. Roughly, these headings also denote the various chapters in the book; however, there is also a particularly well written chapter entitled Resources for Research in MSE, covering federal and industrial research funding, federal laboratories and major national facilities (such as synchrotron and neutron sources). Finally, the book has five long Appendices entitled Synthesis, Processing, Performance (in relation to synthesis and processing], Instrumentation and Analysis and Modeling: these Appendices expand, with specific illustrations, the five themes which the Report as a whole highlights for specialemphasis. It would be fair to say that, among these five, Synthesis and Processing are picked out repeatedly in the text as being crisis areas in the United States, especially in MSE departments in universities which, it is claimed, badly neglect these vital skills. The emphasis on these two skills runs like a refrain right through the text. The Appendices are just as valuable as the main text (if not more so), but one could have wished for more circumstantial case histories, of the type found in the earlier COSMAT Report. As the Harvard Business School long ago recognized, there is nothing like a detailed case history to carry conviction when a thesis is being promulgated. As it is, many of the illustrative 'cases" are hardly more than catalogues, for lack of space. But it is possible that case histories, while fascinating for scientists, would bore politicians!

The chapter on international comparisons is quite convincing to an overseas reviewer and is full of worthwhile statistics and conclusions. In the light of the emphasis, amounting to obsession, of successive MSE committees with (American) National Goals, it is curious to find in this chapter a plaint that the United States cannot match Japan in the systematic formulation of such goals. A faulty machinery for reaching consensus in the United States is highlighted as a deficiency to be put right. Yet many would opine that it is the highly devolved decision-making in the United States and also in Britain (which is here described, in contrast to the highly centralized decision–making in France and Japan) that leads to initiatives and originality: certainly few would claim that MSE research in France is superior overall to that in the 'Anglo-Saxon' countries!

One aspect which is wholly lacking in the book — surprising in a text so clearly targeted at the Washington Establishment — is that of high finance. In comparing the United States and Japan, a book such as *Trading Places — How We Allowed Japan to Take the Lead*, by Clyde Prestowitz (New York, Basic Books, 1988) is a useful corrective. Here we see why Japanese firms and even universities can afford to take a long view while (as the Academy Press book insists) in America, firms and universities are constrained by short–term funding to ensure quick results. Prestowitz documents in painstaking detail how the Japanese government ensures that its favored contractors need take no real financial risks. It might have helped one of the book's clear objectives, to persuade Washington to reform both the magnitude and the nature of federal funding, to venture into a discussion of such matters. Any reviewer can pick holes in particular features of a book. Taking it as a whole and remembering that severe conciseness was clearly aimed at and that the book is unmistakably targeted at a restricted readership, it achieves its objectives well.

97

Supereverything

Contemporary Physics, **31**, 136 (1990)

Superalloys, Supercomposites and Superceramics.. Edited by John K. Tien and Thomas Caulfield. xxvii + 755pp. Boston: Academic. 1989. $ 129.95, ISBN 0 12 690845 1.

My favourite concise dictionary of the English language, Chambers's, defines *physics* as "orig. natural science in general: now, the science of the properties (other than chemical) of matter and energy". This definition, breathtaking in its sweep, is consistent with Rutherford's immortal assertion that "there's physics... and there's stamp-collecting". Chemistry, presumably, rates as stamp-collecting, and so, I suspect, does metallurgy with its later congeners, materials science and materials engineering. Rutherford is dead and gone, and today, perhaps, we can modestly assert that there's physics... and there's philately. For an Englishman, at any rate, one test of the respectability of a 'mystery' is whether its experienced practitioner can write 'Fellow of the Royal Institute/Society of..." after his name. On that basis, philatelists score over physicists, and metallurgists may modestly aspire to parity of esteem with physicists!

The occasion for this bout of lexicographic introspection is the surprising fact that the book under review was sent to *Contemporary Physics* for review, though it has very little indeed to do with physics as I understand it. Perhaps the publishers reckoned, with Terence, that any self-respecting physicist would glance at the book and respond: "Homo sum; humani nil a me alienum puto".

'Superalloys' is an accepted term for the family of alloys used primarily for the hot parts of jet engines; the invented terms 'supercomposites' and 'superceramics' are to be regarded as optimistic echoes of the older word, expressing the hope that the modern versions of these materials will soon overtake superalloys as the preferred stuffs of construction for hot, load-bearing engine and airframe components. There is also a chapter on intermetallic compounds, the probable superalloys of the future: these should, logically, be called $(\text{super})^2$alloys. The story presented here is remarkable, certainly a source of justifiable pride to any materials engineer who reads it.

The centre of gravity of the book lies in *processing* ...the art of getting from the raw materials to the sophisticated end–product...and several chapters (Thermo-mechanical Processing of Superalloys, Powder Metallurgy and Oxide Dispersion Processing of Superalloys, Structural Ceramics: Processing and Properties) are explicitly devoted to this theme. Processing constitutes the heartland of the materials engineer's mystery, and it is made up of a heady mix of physical metallurgy or ceramics, chemistry, chemical engineering, mechanical engineering and control engineering. This can be appreciated when one considers the meaning of 'thermomechanical processing': applied to an alloy, the term refers to the mechanical–cum–thermal sequence of treatments by means of which, say, a loose powder is converted into an elaborately shaped turbine blade. Mechanical engineering comes into the design via the strength and power of the shaping machinery; the thermal history involves the subsidiary mode of expertise known as heat–treatment, which governs the phase structure of the final alloy and is of course intimately related to phase diagrams (which are in the domain of physical metallurgy); chemistry and chemical engineering come into control of impurities, dopants and gaseous environment during processing, which often governs the brittleness or ductility of the end–product; control engineering comes into the automatic control of the whole process, for instance, in the continuous annealing of rolled steel sheet which would be unthinkable without on-line computer control. The heavy emphasis placed in this book on the vital role of vacuum induction melting in making possible advanced superalloys illustrates the point I am making here.

All this said, the book includes several chapters which unmistakably include physics: there is a fine chapter by three authors including one of the editors, Tien, on the theoretical modelling of ternary phase equilibria by the cluster variation method; the chapter on single-crystal superalloys (for turbine blades) includes an application of elasticity theory to compute the elastic characteristics of a crystal in terms of its orientation; there is repeated reference, passim, to the important PHACOMP method of predicting the genesis of deleterious phases such as sigma phases in novel superalloy compositions. Whether one attributes such important topics to physics or to physical metallurgy is of no consequence!

The book comprises 22 chapters by authors (the majority in industry) most of whom work in America, though two (on long-term creep and stress-rupture, and on life prediction and fatigue) are from Japan and one chapter (on carbon-carbon composites) comes from Taiwan. Europe features nowhere among the authors (though at least two of the American authors were born in Britain), and the historical outline of the development of superalloys offered here achieves the remarkable feat of not referring at all to the *Nimonic* family of alloys, developed in Birmingham during and soon after the Second World War... the direct ancestors of today's range of superalloys.

As an experienced editor, I know all too well how hard (nay, impossible) it is to get several dozen authors to finish their contributions on schedule, and this shows up here in the fact that some chapters have no references later than 1984 or 1985... in one extreme case, the most recent references are to 1981! Realistically, the editors cannot be blamed for this; rather they deserve our admiration for getting such a thorough and varied overview into print at all. There are numerous books on superalloys alone, rather fewer books on modern structural (as opposed to electrical) ceramics; fewer books still on composites and probably none as yet on metal-matrix composites. No other book covers the entire gamut of actual and potential materials for high-temperature structural use, and at a uniformly high level of treatment at that, and the editors deserve the gratitude of the materials community for their judgment and determination in bringing this project to a successful conclusion.

98

The Materials Revolution

Advanced Materials, **2**, 440 (1990)

The Materials Revolution. Edited by T. Forester. xiii + 397 pp. Oxford: Basil Blackwell. 1988. Pbk., £ 14.95. ISBN 0-631-16701-3.

Rudyard Kipling laid down the rule that

There are nine-and-sixty ways
Of constructing tribal lays,
And every single one of them is right.

What Kipling omitted at the time to add is that there are four ways of constructing a non-fiction book, and it is an open question whether every single one of them is right. The book can be written by an author (or authors); it can be written by one author and ghosted by another; it can be put together by an editor who commissions various authors to write chapters; and it can be assembled by another kind of editor who collects and reprints, from a variety of books and periodicals, pieces which had been written and published by various authors before he started work himself.

The book under review here belongs to the fourth category. The editor, Tom Forester, was educated (if my memories of Sussex University in the seventies serve me right) in sociology and student radicalism, and has now, it seems, switched his allegiance to science and technology (and the sociology and economics thereof), teaching these topics at Griffith University in Australia. The book contains 20 chapters, grouped loosely under the headings of high–temperature superconductors, materials and society, materials and the economy, materials innovation and substitution, and new frontiers in materials (such topics as processing in space and seabed materials). Most of the chapters are taken from American newspapers, magazines and journals; several of the best contributions are taken from MIT's excellent *Technology Review* and indeed, from evidence in the preface, it appears that the book may have been first published by MIT press, which holds the copyright. There is also one article from a venerable archival scientific periodical, *Science* , and others from a periodical which I had not previously encountered, *The International Journal of Materials and Product Technology.* Some of the authors are highly respected materials science professionals, notably M. C. Flemings, E. D. Hondros, W. D. Kingery and C. S. Smith; the dates of original publication range from 1979 to 1987, with a mean around 1985. Each article has a brief introductory comment from the editor, who also provides a general introduction.

On high-temperature superconductors, we are treated to a passage of hype from *Time* , followed by a sober antidote from *The Economist* and *The Boston Globe.* The editor specifically states that he is counterbalancing hyperbole with sobriety. There is a good deal of journalistic hype throughout the book, in fact, and it carries with it some highly coloured prose, such as: "... gallium arsenide is leaping from the high–tech doghouse to the high–tech penthouse"; this comes from a piece entitled "What's sexier and speedier than silicon?". Other gems include: "Tightening a cable here or lifting a mast there causes mathematical pandemonium" (of fabric roofs), or "The Consequences seemed large, then huge, then mind–boggling; some were wonderful, some were terrifying" (this of molecular biology). In spite of the heightened journalistic language in many places, by and large the treatments are moderately sober. This however is emphatically not true of the last chapter, which the editor admits to have elements of science fiction; it claims to be about "nanotechnology". The piece, which is not about nanocrystalline materials, as might have been supposed in the context of the book, but instead about the prospect of self-assembling and self-replicating entities in biotechnology, materials and information technology, surprisingly comes from an MIT Study Group; MIT is indeed a broad–minded institution!

Not surprisingly, in view of the way the volume was put together, there is much overlap, and some contradiction; for instance, engineering ceramics are extensively discussed (not least in an almost hagiographic chapter about Japanese materials technology, written by Gene Gregory, a — presumably American — professor at a Japanese university), and one is left rather confused about the relative prospects of silicon carbide, silicon nitride and sialon. The editor's lack of scientific background peeps out in his Introduction: "Sialon... is tough enough to withstand almost anything"... which is more, perhaps, than can be said for the reader. There are a number of distinct perspectives on the linked themes of security of materials supply, change of demand with time, increasing efficacy of materials use and the scope for materials substitution: indeed, I think it is fair to say that this theme, with related discussions of the economies of materials, is the most illuminatingly treated part of the book.

It is clear that this book will have only limited appeal for materials professionals (though the supply/demand/substitution part is worth while for such readers) and is in any case largely aimed at a more general readership in industry and commerce. As such, it rates two cheers.

99

Mostly About Aluminium

Nature, 350, 118 (1991)

R & D for Industry — A Century of Technical Innovation at Alcoa.. By Margaret B. W. Graham and Bettye H. Pruitt. 645 pp. Cambridge University Press. 1991. $ 49.50, £ 50.00.

Some years ago, the *New Yorker* carried one of its 'cocktail party cartoons': a group of stolid businessmen is conversing, and one declares: "I've learnt a lot in 62 years, but unfortunately most of it is about aluminum." Drs. Graham and Pruitt have written a massive book that on the face of it is also mostly about alumin(i)um — but it is also about much more. The authors (one of whom is professor of operations management at Boston University) are associated with the Winthrop Group, a company which specializes in the history of business and technology. The book, in the guise of an exhaustive history of the research laboratory of the Aluminum Company of America (Alcoa), is really about the permanent, crucial and insoluble problem of the proper relationship between commercial management and research management in a large industrial company. As the authors point out, aluminium as an industrial material, and corporate industrial research and development are both about a century old; their crises and triumphs developed in synchrony.

Although the German chemical industry already had large research laboratories in the 1880s, it is generally accepted that the first true corporate industrial R & D laboratory in America was set up in a Schenectady barn in 1900 by Willis R. Whitney for the General Electric Company (see the splendid biography *Willis R. Whitney, General Electric and the Origins of US Industrial Research* by George Wise, Columbia University Press, 1985); before that, there had only been troubleshooting laboratories attached to individual factories. After 1900, the idea of segregating corporate researchers in a separate location, sheltered to some degree from day-to-day operating crises, rapidly gained ground. Dupont already had its experimental station in 1902, and its history has been exhaustively analysed in another excellent book recently published by the Cambridge University Press (*Science and Corporate Strategy, Du Pont R&D, 1902--1980,* by David A. Hounshell and John K. Smith, 1988). Most of the points made about that very different company find their echoes in the new book about Alcoa.

Alcoa's business was built on the patent granted in 1889 to Charles Martin Hall, the co–inventor (with the Frenchman Paul Heroult) of the electrolytic process still used today for smelting aluminium. Hall became technical director of the fledgling company, a position which he used to run a small group of subordinates within a factory, under his close direction. He defended his personal authority with tigerish determination when he refused, in 1909, to permit a corporate research laboratory to be established, a determination which he maintained even when he was dying of leukaemia some years later. In consequence, the laboratory was established only in 1919, and the company suffered greatly from this late decision. The battle of Hall versus the rest of the board of directors was a precursor of many later battles about the size, ethos and degree of independence of the company's successive R & D laboratories, battles which swung first one way and then another, culminating in the construction of the Alcoa Technical Center, where research (development moved in earlier) began in 1972.

The later part of the book recounts the decision in the 1980s to recast research strategy in a very radical way, partly as a result of five years' pressure by an outsider brought in to advise on R & D policy. Long–range research was instituted to give the company completely new options to diversify, in due course, from aluminium, greatly increasing total R&D expenditure and also the number of patents taken out; the number of staff with doctorates was doubled, which of necessity led to increased involvement of staff in the activities of their professional bodies, and obliged Alcoa to institute an in–house system of awards to distinguished researchers. In spite of the success of the new R & D strategy, there was further boardroom anxiety about the cost, leading in 1987 to the replacement, by an outsider, of the company chairman who had masterminded the strategy. The creative tension between the commercial ethos and the research ethos, with right on both sides, can never be finally resolved; the current problems at the Philips Research Laboratories show just how difficult this tension is.

This is a profound and stimulating book, always analytical, never hagiographic, rarely censorious. It is a distinguished part of the new wave of industrial history, and well worth the attention of any applied scientist, irrespective of his speciality.

100

A Fox With Quills

Nature, 351, 531 (1991)

<u>*Sir Charles Frank, OBE, FRS: An Eightieth Birthday Tribute.*</u> Edited by R.G. Chambers, J.E. Enderby, A. Keller, A.R. Lang & J.W. Steeds. 448 pp. Bristol: Hilger. 1991. £ 27.50.

Many years ago, Isaiah Berlin, the historian of ideas, in seeking to come to grips with Leo Tolstoy's singular view of history, began by citing a line from an ancient Greek poet, Archilochus: "The fox knows many things, but the hedgehog knows one big thing." (*The Hedgehog and the Fox,* by Isaiah Berlin, London: Weidenfeld and Nicolson, 1953). "The hedgehog", Berlin wrote, "relates everything to a central vision", while "the fox pursues many ends, often unrelated and even contradictory." In analysing the mass of fruitful contradictions that was Tolstoy, Berlin proposed that "Tolstoy was by nature a fox, but believed in being a hedgehog." The inspirer of the book under review here, Sir Charles Frank, began his professional life, perhaps, as a natural hedgehog, but his unquenchable curiosity turned him by degrees into what he is now, a fox furnished with quills.

Frank worked at the Physics Department at the University of Bristol from 1946 until his retirement in 1976, and has subsequently continued to play an active part in bristolian physics. A steady flow of wondrous physics has emerged from that Department during the past half century, and Charles Frank has been the fount, direct or indirect, of much of that. The book under review incorporates 22 essays by eminent scientists in four continents — no fewer than ten of them Fellows of the Royal Society — who either worked under his influence at Bristol or were elsewhere influenced by him during one of his many expeditions to overseas laboratories. The topics, taken in conjunction, eloquently illustrate a remark, cited here, which Charles Frank made in his retirement speech in 1976: " Physics is not just concerning the nature of things, but concerning the interconnectedness of all the natures of things."

Scientific subjects covered by the essays include growth and morphology of crystals; interfacial structure, notably in epitaxial overgrowths; defect–induced pyroelectricity; liquid crystals; incommensurate and quasi–crystals; stability of and defects in diamonds; polymer crystallization; creation of strong polymeric fibres by liquid- or solid-state processing; electrolytic 'parting' of gold alloys; the role of material flow in geophysics, especially in plate tectonics; several essays dealing with abstruse sidetracks of dislocation theory; and a remarkable one by Ponomarev of Moscow on muon–catalysed nuclear fusion, a flourishing research field directly initiated by one of Charles Frank's short but dense notes, dated 1947, itself occasioned by a mysterious observation by the Bristol cosmic ray physicist, C.F. Powell, which caught Charles Frank's roving attention, no doubt during laboratory tea. Indeed, there are many asides in this book about the fruitfulness of teatime conversations; a systematic study of the central role, in stimulating academic research, of laboratory tea, a largely British institution, is overdue. (I have much personal experience of the impact that this underrated institution has on foreign visitors to British universities!)

The most eloquent account of Frank in action at Bristol is to be found in the splendid essay by Andrew Keller on chain-folded crystallization of polymers. The essay depicts the Bristol laboratory as it first struck a brilliant, emotional young Hungarian chemist: "Physics was in the air, was discussed everywhere: on the stairs, over tea, in the doors...; passage of time was simply forgotten or ignored. There was no distinction between high and low brow, it was all an intellectual adventure. That is how polymers eventually started in between quantum mechanics, particle physics, liquid helium, design of optical instruments, and much else. Here I saw science in action, not fragmented into specialities but as an indivisible whole, a single enterprise of the human mind.... And Sir Charles was central to it all! Like a chess virtuoso playing several games simultaneously he was conducting these unforgettable teatime discussions on virtually all subjects in science."

The discovery of chain-folded crystallization of polymers, actually polyethylene, was a truly remarkable episode — wholly unexpected, *a priori* wildly improbable, as indeed were a number of other observations also recounted in the book which owed a good deal to Frank's stimulation. Keller's findings led to fierce and sustained controversy, only now gradually diminishing, as to the mode of formation of chain-folded crystals. His account, besides being a valuable historical record, also furnishes a sovereign contradiction of those deluded critics of the scientific method — we have all encountered them — who accuse its practitioners of an arid and passionless pursuit of bare, unaccommodated fact.

Some of Frank's ideas relate to the work of nineteenth-century physicists, such as his very recent work on the vectorial depiction of the rotational relationships of neighbouring crystal grains (the Frank-Rodrigues method, here outlined by Robert Pond); he often found inspiration in physics of the distant past, in several languages, which is no doubt why Frank Nabarro includes a stimulating essay, full of 19th-century quotations in French and German, about a curiosity in crystal physics involving pyroelectricity in crystals which on the face of it should not exhibit it. Reverting to the Frank-Rodrigues approach, it is characteristic of Charles Frank that the driving force for this work was his recollection of a practical problem put to him during a consulting visit to a U.S. laboratory, over 30 years ago!

The first essay in the book, written by R.V. Jones with his inimitable brand of enthusiasm, describes Frank's student days and his subsequent war-work in intelligence. In the episode of his successful interpretation of a mysterious air reconnaisance photograph, his natural propensity to think in geometric terms is made evident. (A love of geometry is what made the early Frank a hedgehog.) This propensity is also consistent with his early concentration on dislocations, archgeometrical entities, which led in 1949 to his best-known prediction, that of dislocation-mediated crystal growth, unforgettably confirmed by an experimentalist minutes after the theoretical case had been publicly expounded by Frank. (The prediction itself resulted from a mismatch between calculated and observed supersaturations for rapid crystal growth, an example of his habitual reliance on rigorous quantitative arguments). The concept of the nonclosing Burgers circuit around a dislocation line motivated Michael Berry, in another splendid essay, into generalising this notion into many other kinds of nonclosing circuits in several kinds of space, not necessarily 'real'.

From Berry's very abstruse piece of theory it is a long step to Kenneth Ashbee's account of bidimensional compression (originally proposed by Charles Frank with chain alignment of polymers in mind), now useful *inter multa alia* for extruding cider from apples, and to Donald Hurle's application of Charles Frank's ideas on crystal morphology (elsewhere expounded in a lucid essay by P. Bennema) to the suppression of growth twins in semiconductor crystals. All is grist, or apples, that comes to the mill.

Every physical scientist will find something here to fascinate him. As a tribute to a great, exceedingly influential physicist, this volume could not be bettered.

Index

(This list cites serial numbers of articles, *not* page numbers)